Critical Criminological Perspectives

Series Editors
Reece Walters
Faculty of Law
Queensland University of Technology
Brisbane, QLD, Australia

Deborah H. Drake
Department of Social Policy and Criminology
The Open University
Milton Keynes, UK

The Palgrave Critical Criminological Perspectives book series aims to showcase the importance of critical criminological thinking when examining problems of crime, social harm and criminal and social justice. Critical perspectives have been instrumental in creating new research agendas and areas of criminological interest. By challenging state defined concepts of crime and rejecting positive analyses of criminality, critical criminological approaches continually push the boundaries and scope of criminology, creating new areas of focus and developing new ways of thinking about, and responding to, issues of social concern at local, national and global levels. Recent years have witnessed a flourishing of critical criminological narratives and this series seeks to capture the original and innovative ways that these discourses are engaging with contemporary issues of crime and justice.

More information about this series at
http://www.palgrave.com/gp/series/14932

Diego Canciani

The Politics and Practice of Occupational Health and Safety Law Enforcement

palgrave
macmillan

Diego Canciani
Department of Social Sciences
University of Roehampton
London, UK

Critical Criminological Perspectives
ISBN 978-3-319-98508-4   ISBN 978-3-319-98509-1   (eBook)
https://doi.org/10.1007/978-3-319-98509-1

Library of Congress Control Number: 2018950561

This Palgrave Macmillan imprint is published by the registered company Springer Nature Switzerland AG
The registered company address is: Gewerbestrasse 11, 6330 Cham, Switzerland

This book is dedicated to my parents, Mariagrazia and Adriano

Acknowledgements

I would like to thank my colleagues and friends at Roehampton, London South Bank and at the European Group and beyond, whose useful suggestions and encouragement have helped me throughout every stage of the preparation of this book.

I would also like to thank my Ph.D. supervisors, whose incredibly useful suggestions and support have allowed me to complete successfully my Ph.D., the work of which this book is based on.

I would also like to thank all my family members whose continuous support, encouragements, discussions and love have greatly helped me to always believe in myself and strengthen my resilience.

Finally, I would like to thank my friends, who have given me great times and hence helped me to forget about this book when that was necessary.

Contents

Abbreviations

ASL	Local Health Agency/Azienda Sanitaria Locale
BRE	Better Regulation Executive
CBA	Cost-Benefit Analysis
CC	Civil Code/Codice Civile
CCA	Centre for Corporate Accountability
CP	Penal Code/Codice Penale
CPC	Civil Code Procedure/Codice di Procedura Civile
CPP	Code of Penal Procedure/Codice di Procedura Penale
CPS	Crown Prosecution Service
CPSR	Permanent Conference between States and Regions and the Autonomous provinces of Trento and Bolzano/ Conferenza Permanente per i Rapporti tra lo Stato le Regioni e le Province Autonome di Trento e Bolzano
DETR	Department of the Environment, Transport and the Regions
DRL	Regional Labour Directorate/Dipartimento Regionale del lavoro
ECJ	European Court of Justice
EEC	European Economic Community
EMM	Enforcement Management Model
EU	European Union

FFI	Fee for Intervention
FOD	Field Operations Directorate
GMB	General trade union
HASW Act 1974	Health and Safety at Work Act 1974
HID	Chemical Industries and Specialist Industries
HSC	Health and Safety Commission
HSE	Health and Safety Executive
ILO	International Labor Organisation
INAIL	National Institute for Insurance against Working Accidents/Istituto Nazionale per l'Assicurazione contro gli Infortuni sul Lavoro/National Institute for Insurance against Working Accidents
INPS	National Institute for Social Security/Istituto Nazionale della Previdenza Sociale
ISI	Business Support Incentives/Incentivi di Sostegno alle Imprese
ISIC	International Standard Industrial Classification
ISPESL	Superior Institute for Prevention and Work Safety/Istituto superiore per la prevenzione e la sicurezza del lavoro
ITL	Labour Territorial Inspectorates (ITL)/Ispettorati territoriali del lavoro
LA	Local Authority
LFS	Labour Force Survey
ND	Nuclear Division
ORR	Office of Rail Regulation (Office of Rail and Road since 2015)
PM	Public Ministry/ Pubblico Ministero
ppp	purchesing power parity
RIDDOR	Reporting of Injuries, Diseases and Dangerous Occurrences Regulations
RoSPA	Royal Society for the Prevention of Accidents
SME	Small and Medium Enterprise
SoSWP	Secretary of State for Work and Pension
SPSAL	Prevention Services for Safety in Work Environments/Istituto di Prevenzione e Sicurezza Ambienti di Lavoro
TUC	Trade Union Congress
UPG	Judiciary Police Official/Ufficiale di Polizia Giudiziaria

List of Figures

List of Tables

Abstract

This book analyses and compares British and Italian occupational health and safety (OHS) enforcement policies. Regulatory enforcement policies is a highly debated field of study in the academic literature and politics. The extended amount of harm and suffering caused by OHS incidents in developed countries and the limited concern that a number of academics and politicians have on this issue, and on the under-criminalisation of these crimes, exposes a fundamental controversial aspect of modern liberal values of social justice and equality. The innovative regulatory enforcement policies theorised by academics and adopted by jurisdictions since the 1970s is evidence of this fundamental issue and a transnational comparative analysis of jurisdictions with different social, political and criminal justice system values offers an important contribution to the literature and political debate. This book compares British and Italian OHS enforcement policies in a Marxist critical theoretical framework by analysing how modern socio-political values and globally accepted means of economic production embraced by developed jurisdictions are the root causes of the under-criminalisation of OHS crimes. The book analyses the British and Italian legal and political values and practices, the historical contexts in which the regulations and enforcement policies developed and compares OHS incidents trends longitudinally. It also analyses how OHS enforcement institutions are funded and politically controlled and the fundamental problem with law enforcement discretionary practices, and how these affect the fundamental modern principles of legal fairness and consistency. This book exposes the embedded inequality of developed nation-states socio-political and criminal justice systems and how global political and economic contemporary trends, together with the ever-increasing challenges of preserving liberal values, are gradually eroding the legitimacy of modern nation states and their claimed fundamental commitment to social justice and equality.

1

Introduction

Occupational safety incidents[1] and occupational health-related sicknesses are the most common causes of suffering in developing countries. Yet, occupational health and safety (OHS) crimes in Britain[2] are subject to

---

[1]Throughout the thesis I use the term incidents rather than accidents because I agree with Tombs and Whyte (2007) who "view safety crimes as crimes of violence. Though this is not an original approach, it is, as we shall see, one which, in the context of academic criminology and general representations of occupational injuries, a rarity. For us [Tombs and Whyte], this lack of criminological attention to safety crimes as crimes of violence is less a quality of the latter phenomena, more a failure of the discipline to reflect upon long-standing, but ontologically weak, assumptions. Thus, once occupational injuries are viewed not as accidents but as incidents which are not only largely preventable, but which the law requires to be prevented, then they fall within the ambit of criminology. Then, if we consider these illegalities in terms of their potential or actual consequences - injury and death - we realise that these look remarkably similar to the results of those events that most men and women as well as policy-makers, politicians, and academics, deem to be 'proper' violence. Most crucially, these latter conceptions of violence are [...] based upon an implicit association of violence with the inter-personal and with intention, both qualities that are often absent from safety crimes. That safety crimes are crimes of (actual or potential) violence is, for us, unequivocal; yet to reach this conclusion one must generally move beyond criminology, to understandings of violence developed in other disciplines" (Tombs and Whyte 2007, pp. 5–6).

[2]This research refers to the British occupational health and safety enforcement institutions because the HSE has jurisdiction is England, Wales and Scotland. Hence, referring to the English and Wales OHS enforcement institution would be wrong. However, the reader should be aware that the Scottish political and legal system, due to devolution occurred from the British parliament in recent years, is different from the English one, and arguably adopting a criminal justice system and

© The Author(s) 2019
D. Canciani, *The Politics and Practice of Occupational Health
and Safety Law Enforcement,* Critical Criminological Perspectives,
https://doi.org/10.1007/978-3-319-98509-1_1

chronic form of under-criminalisation, which means violent injuries and sicknesses resulting from working conditions have traditionally been punished significantly less frequently and more leniently than other similarly harmful crimes (Carson 1970a, b, 1979; Fooks 2008; CCA 2008). Given that employers have an absolute responsibility over the health and safety of employees and that it is estimated that there are 12,000 deaths caused by occupational incidents and occupational diseases per year in Britain alone. We should expect the prison population to be mainly composed of employers and company directors, but this is not the case. Firms, employers and company directors—especially of larger firms and corporations—rarely go to prison and are often inadequately penalised (Reiman and Leighton 2010). In addition, British employers, corporate directors and firms pay only a small part of the social costs caused by OHS incidents, such as medical costs, lost working days and invalidities (CCA 2008; Fooks 2008; Carson 1970a, b, 1979). In fact, the Health and Safety Executive (HSE), the main British OHS regulator, has published a report estimating that between 2006 and 2011 employees and their families, on average, bore 60% of the cost of OHS incidents, while employers and the state bore 20% each (HSE 2013a). The findings from this study suggest that in Italy the level of redistribution seems better due to a national compulsory OHS insurance, but safety crimes appear to be equally under-criminalisation. However, while the under-criminalisation in Britain results from an inadequate reaction of the regulatory institutions, in Italy this happens during court trials. This comparative research study investigates why this is the case.

The scope of this book, thus, is to shed light on the reasons for the under-criminalisation of OHS crimes in Britain and Italy by comparing the enforcement policies and criminal procedures enforcement institutions must follow. Cross-national comparative analyses are challenging and the differences between these two countries' social, historical and

---

political model resembling the continental due-process non-adversarial one more. In addition, the fieldwork informing this comparative analysis has been conducted in England and thus the study omits to analyse and compare the opinions of Scottish occupational health and safety enforcement officers.

political traditions can barely be accounted for in this book, but as Nelken (2010) argues, these differences are exactly the reasons why these comparisons are worth pursuing. This comparative study aims to scrutinise the policies governing the enforcement institutions by also taking into account the political processes determining these policies and other global political and economic trends that affect these decisions. The remaining part of this chapter aims to introduce the subject by considering the rationales for comparing these two jurisdictions and exploring the fundamental differences between these two countries that are relative in understanding how and why OHS crimes are under-criminalised.

The paths in policy development that have affected these two countries' OHS enforcement practices are very relevant for the literature debate and social policymakers. The 1970s heralded the historical period when goal-setting philosophies started to be introduced for OHS legislations (Lordo Robens 1972). In 1989, the European member states agreed to follow the European Economic Community (EEC) Council Directive (89/391/EEC) to harmonise OHS regulatory frameworks in order to eliminate economic competitive advantages in the single market (Council of the European Union 1989). The EEC Council Directive on occupational safety and health (89/391/EEC) instituted a minimum OHS standard to implement across member states. The adoption of OHS goal-setting legislation and the changes to the nature of economies' production that has been happening in Europe since the 1970s has occurred vis-à-vis the implementation of innovative law enforcement methods, which are based more on the use of education and support than on deterrence (Pearce and Tombs 1990, 1991, 1998). The British and Italian reforms, however, did not follow similar paths and from the middle to the end of the 1990s started to diverge due to the fundamental legal differences of these two criminal justice systems.

A legal rationale for comparing these two countries is that each has complied with the 1989 EEC Council Directive on OHS (89/391/EEC). In fact, by the early 1990s OHS legislation in Europe was harmonised and countries like Italy had also implemented OHS goal-setting regulation. Although this legislation has been designed mainly to ensure a fair competition across member states, the policies adopted to enforce the regulations have not been harmonised and today might remain one of the most important areas that European governments can act on to create an economic competitive advantage against other European Union (EU) member states. Therefore, these disparities create inequalities for businesses, workers and citizens within jurisdictions and across Europe that are interesting to study in their own right and might become subject to future European-wide regulations. The enforcement policies adopted might also be considered as even more important than the actual regulations, because these will have a significant effect on the levels of compliance. In other words, a regulation might just be a formality if it is not enforced.

Hence, the practices used to enforce OHS regulation are key to ensure compliance, to prevent harm and achieve social justice and equality. There has been much discussion about the most effective enforcement policies to adopt in order to ensure compliance. A comparative analysis of two countries with different OHS enforcement policies but similar regulations, such are Britain and Italy, offers a significant contribution to the literature on the subject. The academic literature discussing this topic, however, is fundamentally divided by the theoretical framework adopted to analyse the issue.

1   Crimes of the Powerful and the State

This research study uses a critical Marxist theoretical framework to understand the causes of the under-criminalisation of OHS crimes in both Britain and Italy. Pearce (1976) has been one of the first to recognise and analyse the social discrimination caused by the different reactions of the criminal justice system from a critical Marxist perspective. He argues that criminologists' study of crime and the criminal justice system's responses to crime have often been constructed with a labelling theoretical perspective. These assume that the criminal justice system is responsible for labelling the working class as deviant, either as a response to their actions, or, more critically, as a result of the criminalising actions of the criminal justice system itself. The latter process, Pearce argues, occurs because the modern state is directly conditioned and subtly controlled by the bourgeoise class and capitalist interests. As a consequence, OHS legal breaches are under-criminalised because the regulation represents a limit to the economic profits and political power of the bourgeoise class.

Orthodox labelling theorists do not critically analyse how power is structurally distributed throughout the state and law enforcement institutions. They take for granted that power is evenly distributed across social individuals, or factions, and uncritically presume that social decisions are reached consensually through political compromises. Lemert (1967 and 1972), Pearce (1976) argues, criticises Sutherland (1949) for arguing that corporate crime represents the sine qua non of capitalist societies, but instead he sees the deviance of the powerful and elite social entities simply and uncritically as a pragmatic response to pragmatic social problems. A Marxist analysis capable of recognising class struggles is, thus, essential to understanding the causes of the under-criminalisation of OHS crimes in both Britain and Italy.

Labelling theorists fail to explain critically the reasons for the uneven distribution of criminalisation in society. The over-representation of the working class in the criminal justice system, courts and prisons is caused by these social classes' lack of political organisation and representation (Pearce 1976). They are not failing to organise politically because they have no will or capacity to do so, but because bourgeois interests constantly undermine political organisations and actions that can represent their interests. Labelling theorists, thus, do not question the reasons for the criminal justice system apparatus's over-response to lower social class crimes and so tend to ignore the under-criminalisation of OHS crimes when theorising new enforcement policies. This argument, however, questions what the role of the state—or any governing institution perhaps—is and should be. This issue demonstrates how the importance of the survival of the modern bourgeois state is counterposed to ideas of social justice and equality.

Marxists see the state as a modern capitalist institution underpinned by bourgeois interests. The survival of the modern capitalist state can only be ensured through a form of foreign and national security actions. The former is dealt with by the military industry, which is funded by the surplus value created through a capitalist economy. The latter occurs through the maintenance of a seemingly consensual capitalist society, through the use of subtle political manoeuvres, or through the outright use of force, especially when policing dissenting social groups, to weaken and eventually eradicate the development of any social class consciousness that is not aligned with bourgeois capitalist values. The state and the key institutions composing it are, thus, designed to both support capitalist interests, but also legitimise their actions and decisions under modern liberal values of justice and equality. Pearce (1976) argues that the state's support for capitalist interests is created through forms of capitalist-led social consciousness and used to fabricate political consensus. The dissent emerging from this false capitalist-led consciousness, however, is one of the liberal problems because bourgeois class interests are more important than liberal values of justice and equality.

Thinking of the under-criminalisation of OHS crime in redistributive justice terms is useful. Rawls (1973, pp. 575–576 as cited in Cowling, 2011) argues that crimes upset the pattern of social redistribution, which means that criminal justice system institutions' scopes should be to rebalance justice. This becomes very clear in the context of regulatory crimes because as Taylor (1972 as cited by Ogus 1994) argues, OHS regulations developed in Europe as a pragmatic response to social problems caused by the industrialisation of economies. In other words, economic regulation ensures a redistribution of rights and wealth from the bourgeoisie to the working class. Contemporary nation-states claiming to embrace liberal values, such as Britain and Italy, have the capacity to limit the inequalities and injustices that capitalism creates through laws and regulations, but these are a direct threat to the expansion and survival of the state and the capitalist means of production. Human rights laws, modern constitutional laws, the separation of powers doctrine and criminal justice system due process procedures are all there to limit power imbalances and promote modern liberal values. The involvement of the state in these affairs has decreased significantly since the 1980s when developed counties started to adopt Neoliberal economic policies. Neoliberal economists suggest that state involvement creates even more inequality and that citizens should be given more responsibility for their failures. However, the under-criminalisation of OHS crimes demonstrates that the bourgeoisie also seems to avoid taking responsibility for the social harm they cause. In other words, Neoliberals argue that state involvement in economic policies is detrimental, but do not question and criticise the inequalities created by Neoliberal laissez-faire economic policies. So, less state involvement means that its redistributive power to promote liberal social values is undermined. When the state decides to decrease the level of regulatory enforcement policies, modern liberal political thoughts of social justice and equality before the law are undermined.

Understanding the fundamental differences between the British and Italian procedures dictating law and regulatory enforcement activities and criminal procedures helps to recognise nation-states' roles in the under-criminalisation of OHS crimes. Comparing the British and Italian legal systems' traditions, such as the differences between the crime control model and the due process model, and the differences between the adversarial and non-adversarial court trial system, as well as how the separation of powers doctrine can influence political and criminal justice system institutions becomes key in a critical analysis of OHS enforcement policies.

2 British and Italian Legal Systems

The British and Italian criminal justice systems are fundamentally different. While the British legal system has traditionally been associated with an adversarial and crime control criminal justice model, the Italian system has been associated with a non-adversarial (or inquisitorial)[3] and due process model.[4] With the exclusion of Scotland,[5] the policies used

---

[3]The inquisitorial name derives from the religion inquisitions which characterised justice during the middle ages, but this term is slightly misleading in this context. Hence, in this book I will adopt the term non-adversarial, rather than inquisitorial.

[4]It is important to note that these models are different in *principle*, which means that policies and practices can, at times, be quite similar across jurisdictions. Only the principles and policies that are relevant to this research study will be analysed.

[5]See Footnote 2 above.

to enforce the OHS regulations in England and Wales and Italy are significantly influenced by these two different legal traditions. It is crucial, therefore, to position the OHS enforcement policies, and the theories underpinning them, within the broader framework of the criminal justice systems of these two jurisdictions. These differences will help to frame the comparative analysis of this research study.

Sanders and Young (2007) argue that

> the criminal justice system is [...] a complex social institution which regulates potential, alleged and actual criminal activity within procedural limits supposed to protect people from wrongful treatment and wrongful conviction. (Sanders and Young 2007, p. 1)

The criminal justice system is a complex interrelation of procedures and rules applied by a wealth of institutions and individuals. *Criminal* conduct is, therefore, a socially constructed concept; the achievement of *justice* depends on whether alleged criminals are apprehended, convicted or acquitted; and the *system* consists of the complex interrelation of legal institutions that are given the responsibility to make these decisions (Sanders and Young 2007; Davis et al. 2010).

The primary aims of the criminal justice systems should be the effective and efficient apprehension of criminals and the protection of civil liberties and rights of citizens (Sanders and Young 2007; Packer 1968; Davis et al. 2010). The issue, however, is that the protection of the civil liberties and rights of citizens that have been victims of criminal activities has to be balanced against the civil liberties and rights of citizens suspected of committing criminal activities. The full achievement of these two objectives eliminates miscarriages of justice (Sanders and Young 2007; Davis 2004). In the context of this research study, the under-criminalisation of OHS crimes represents an imbalance of these rights because the employers and organisations committing them are rarely made responsible for their actions, apprehended, penalised and prosecuted and because victims still bear most of the financial and social costs of these crimes.

The institutions responsible for achieving this balance form the criminal justice system (Rogers 2006). The emphasis given to either of the two main aims of the criminal justice system, however, might change across jurisdictions. Law enforcement institutions, prosecution services and courts of justice all play roles that are capable of offsetting this balance. They might have the power to decide when and how much pro-active and reactive enforcement activities[6] to conduct and whether to take breaches to court, such is the case in Britain, or might not, such is the case in Italy. They might be constrained by ambiguous and interpretable legal definitions when attempting to demonstrate the suspects' guilt, such is the case for OHS crimes (Sanders and Young 2007). The procedures used in these contexts have a direct impact on the way civil liberties are preserved in society (Packer 1968; Sanders and Young 2007). Thus, as Packer (1968) defines, the quicker and more accurate a criminal justice system's response to crimes is, the more *efficient* the system will be (Davis et al. 2010; Sanders and Young 2007; Packer 1968). The concept of *legal efficiency* is essential for this study because the under-criminalisation of OHS crimes in Britain and Italy is partially caused by the inefficient responses of the criminal justice systems.

To compare the methods adopted by different jurisdictions to apprehend and convict or acquit suspected criminals, scholars have broadly modelled the criminal justice systems across economically advanced countries along two dimensions. The first is the system adopted to judge suspected criminals, which can be adversarial (accusatorial) or non-adversarial (inquisitorial) (see Damaška 1973). The second is the overall model defining the primary aim and general processes of the justice systems, typified (usually) as fitting either the crime control or the due process model (Damaška 1973; Packer 1968; Ma 2002; Sanders and Young 2007; Sung 2006).

These different categories are usually—often crudely—adopted to compare countries using Common law and Civil law models and has been identified as a significant difference in this research study. While both traditions come from Roman law, the Common law British system is usually associated with the crime control model and adversarial

---

[6]*Proactive* describes enforcement institutions' pre-planned annual enforcement activities, while *reactive* describes enforcement activities that occurs after incidents.

legal systems, which was influenced by eleventh-century Norman invasions. The Civil law Italian system is associated with the due-process model and inquisitorial (non-adversarial) legal systems and influenced by religious law originated in the thirteenth century (Damaśka 1973). Although contemporary legal systems in most jurisdictions are still distinguishable between Common and Civil law systems, scholars acknowledge that these jurisdictions are all *mostly* designed to respect citizens' civil rights, avoid miscarriages of justice and comply with fundamental human rights. Despite these similarities, however, there are still differences that can be analysed in order to improve the understanding of the OHS enforcement policies in Britain and Italy and why OHS crimes are subject to forms of under-criminalisation.

The first significant difference between the British and Italian legal traditions can be explained by reference to the crime-control and due-process models of criminal justice systems. Britain has traditionally adopted the former, while the Italian's is characterised by the latter. Packer's (1968) famous classification of crime-control and due-process procedural models highlights specific features of the criminal justice systems in countries characterised by these two legal traditions (see also Sanders and Young 2007). Two characteristics are particularly relevant for this comparative study.

The underlying values of the crime-control model are based on the primary aim of repressing criminal conduct. Failure to control crimes means that the personal freedom of law-abiding citizens is undermined. According to classical theorists, the criminal justice system represents the primary guarantor of social freedom, and crime-control values require suspect screening, determination of guilt, and appropriate disposition of suspects as quickly and economically as possible (Jones 2013). These assumptions have been criticised by Marxist criminologists who argue that while the aim of repressing the criminal conduct is essential and considered an urgent matter when targeting lower social classes, it is less so when targeting upper social classes (Jones 2013; Reiman and Leighton 2010).

"If the crime-control model resembles an assembly line, the due-process model looks very much like an obstacle course" (Packer 1968, p. 163). In the due-process model, the deprivation of individuals' freedom should be avoided. The due-process model considers the fact-finding process as a formal but non-adjudicative stage and where errors are more likely to happen. It is adopted because witnesses usually produce inaccurate accounts of events, especially for the most disturbing crimes, which means that statements and depositions might be incorrect. Hence, the due-process model prefers a

> formal, adjudicative, adversarial fact-finding process in which the factual case against the accused is publicly heard by an impartial tribunal and is evaluated only after the accused has had a full opportunity to discredit the case against him. (Packer 1968, pp. 163–164)

In the due-process model the fact-finding process must not be eliminated and the adjudicative process should be repeated as many times as necessary. This research study demonstrates that the way OHS crimes are targeted, enforced and prosecuted at times contravene these legal principles.

The financial efficiency of the criminal justice system process is an essential consideration in the crime-control model, but not in the due-process model. For Packer (1968) *efficiency* is "the system's capacity to apprehend, try, convict, and dispose of a high proportion of criminal offenders whose offences become known" (Packer 1968, p. 158). The criminal justice system, however, operates within limited resources, which means that it must achieve a high rate of apprehensions and convictions in a very efficient and effective way, by incrementing the speed with which crimes are dealt with and by reducing the opportunities for the claimant to challenge the process and decisions leading to conviction. This is particularly the case in the crime-control model where the selection of cases for criminal prosecution is not always standardised. Informal procedures in police stations are preferred because court trials are expensive and time-consuming. In the due-process model, the consistency of the legal procedures to follow are much more important than the financial efficiency of the adjudication process. The adoption of the due-process model has usually evolved from the important priority of maximising the preservation of individuals' rights and limiting the unchecked application of official (state) power. For the due-process model, the maximum efficiency and power to investigate crimes means maximum tyranny because unchecked official power would lead to an abuse of citizens' liberties to achieve convictions (Packer 1968). This is a key difference between Britain and Italy because the definition of legal and financial efficiency has, at times, been used interchangeably when analysing the enforcement policies used to police OHS crimes.

The second significant difference between the British and Italian legal traditions can be explained by reference to the adversarial and non-adversarial legal systems. Damaška (1973, 1986) argues that adversarial and non-adversarial procedural systems in contemporary democracies have a number of different characteristics (Sanders and Young 2007). There are fundamental differences, such as the defendant's right to remain silent, the right of the parties to access the available evidence, the defendant's right to plead guilty, and the role of the judge (Damaška 1973, 1986; Sanders and Young 2007). Others, however, are much more important to understand for this comparative analysis.

The primary aims of the two systems are different. The primary aim of the adversarial system is to find the evidence to reach a verdict. It is based on the idea that proceedings should be structured as a dispute between two sides in a position of theoretical equality before a court, which decides on the outcome of the trial. The parties have definite, independent, and conflicting interests and the judge's role is that of a referee. The prosecutor's role is to obtain a conviction and the defence's is to argue against it, or when the defendant is found guilty to provide mitigation with the aim of securing a lesser sentence. Conversely, the non-adversarial system is based on the primary aim of finding the truth. A claimant, such as a public prosecutor, is essential to start the investigation, while the defendant will have the right to participate actively in the trial process and when deciding a suitable sentence. The investigation itself is conducted by a judge, who actively takes part in the reconstruction of the circumstances that led to the alleged criminal conduct. The judge is not a referee, but a proactive and participating member in the trial. The non-adversarial system expects both parties to play a more active role and to cooperate in the collection of evidence than in the adversarial system (Damaška 1973, 1986). This difference is significant in this research study because the complexity of the OHS regulation benefits from cooperation between parties, but while in Britain this cooperation is more welcomed at enforcement stage, in Italy it is only expected to happen in courts of justice.

The assumptions that trigger the beginning of a trial are also different. In the adversarial system the assumption is that the crime has probably been committed and the law enforcement institutions have a degree of discretion when deciding whether to continue pursuing the court prosecution. This means that traditionally there have not been independent institutions responsible for taking these decisions. In recent years, however, prosecution services have started to appear in many adversarial systems to ensure consistent application of the law, such as the case of the Crown Prosecution Service (CPS) in England. In the non-adversarial system, the process is triggered by the initial probability that a crime has been committed and the case is then transferred to an independent adjudicator, such as the Public Ministry (PM) in Italy, which decides whether to start the court trial. However, decisions to dispose of cases are less likely to be subject to discretion, and in Italy it is compulsory if there is enough evidence of criminal conduct but might require a judge's approval. Prosecutors are guided by specific official procedures that aid their decisions and ensure a consistent application of the law (Damaŝka 1973, 1986). Hence, this is a key difference in this comparative analysis because while in Italy prosecutable OHS breaches must be taken to court, in Britain this depends on the discretionary decisions of the enforcement institutions.

In terms of procedural barriers erected to protect the human rights of suspects, Damaŝka (1973) argues that these ensure the correct verdict is achieved, but it is also a barrier for the collection of evidence and, ultimately, the protection of law-abiding citizens. Both the adversarial and non-adversarial systems aim to achieve this goal, but in different ways. The adversarial system is characterised by lower evidential and procedural barriers in order to achieve justice easily and cheaply and ensure public safety quickly. This means that guilty pleas are welcomed and ideal. The enforcement institutions, however, might raise them to ensure that prosecutable cases achieve guilty sentences and hence are legally efficient and resources are used more efficiently. The non-adversarial system might lower evidential barriers because the primary aim is to gather the evidence needed to achieve the truth, but might also raise them because the pre-trial procedures, according to due-process procedures, are also meant to protect the suspect's rights (Sanders and Young 2007; Damaŝka 1986). This aspect is also essential in this comparative study because procedural barriers can not only have different and complex decriminalising effects during the enforcement of OHS regulations, but also during court trials.

The way sentences are decided in both systems also differ. In the adversarial system sentences also depend on characteristics, such as the annual revenue of a firm. The substantive role of law, due to its uncodified origins, does not allow the criminal justice system to issue standardised punishments. Thus, verdicts do not necessarily aim to be proportionate to the harm caused but depend much more on common sense and traditions, or case law. The traditional involvement of juries has tried to repair this issue, but juries are also less likely to understand the technical evidence in complex cases, such as regulatory cases. In the non-adversarial system, the substantive law is codified to ensure consistency, and therefore sentences are much more likely to be proportionate to the harm caused. The non-adversarial system relies much more on criminal codes and procedures than on traditions and past cases. Despite this, the technical evidence of complex cases can also impair the capacity to prosecute for the different charges, and hence for the severity of the sentences (Damaŝka 1973, 1986). Sentences and penalties for OHS crimes in Britain and Italy have traditionally been low. This comparative analysis will aim to explore the reasons behind this issue.

A final fundamental difference between Britain and Italy is the level of political checks and balances—the separation of state powers—implemented across the political institutions, and, in particular, between those responsible for managing the OHS enforcement institutions in Britain and Italy. The separation of powers doctrine has Babylonian origins, developed during the ancient Greek Republican city-states, but it was Montesquieu (de Secondat 2001) who, following ideas developed from John Locke and American and European political debates, reformulated it into a theory that has determined the divisions of powers within modern nation-states (Vile 1963). Montesquieu argues that to avoid despotic rulers, guarantee personal civil liberties and create democratic governments, the state should be formed of three main institutions, each with specific responsibilities; legislating (legislative), executing legislations (executive) and judging legal violations (judiciary). Montesquieu's theory was inspired by Roman law and British constitutional law, which were both based on the tripartite system (Sabine and Thorson 1973). Yet, the British legislature and executive had, and still has, a political relationship between institutions that is much more intertwined than the one envisaged by Montesquieu (de Secondat 2001). This is a significant difference between Britain and Italy and between the common law and statute law traditions. The former has been the traditional legal practice in Britain since the Norman Conquest, while the latter developed from Roman Law, was injected into the Italian peninsula during Napoleon's military campaign and finally well embraced in the 1948 post-war Italian Constitution (Vile 1963; Langbein et al. 2009; Mousourakis 2015). Hence, to ensure civil liberties, equality, democratic principles and avoid civil war, these three institutions should be mutually dependent on each other, but also scrutinise each other operations through processes of checks and balances.

The executive, or government, is responsible for executing laws and initiating reformation processes by proposing new statutes to the legislature, or parliament, for approval (Vile 1963). The execution of laws approved by the legislature must occur within the remit of the laws approved by the legislature, or the executive would abuse its powers. The legislature represents the interests of most social factions and is, thus, responsible for checking the operations of the executive and by considering propositions for new statutes. The legislative has no executive powers but has the responsibility to empower the executive's actions by approving or disapproving new statutes. The vote of no confidence, for example, can be used to test whether the majority of the publicly elected members of parliament are confident in the capability of the government to govern. Montesquieu contributed to the theory by introducing an independent judiciary (de Secondat 2001). The judiciary is independent from the executive and the legislature and has the responsibility of applying the law and resolving social disputes between the state and citizens and between citizens according to the law.

In contemporary democracies, the powers assigned to the executive, legislative and the executive can change (Warwick 2006). For example, in common law countries, the judiciary has an institutionalised power allowing it to interpret the law according to contemporary social tendencies, whereas in statutes law countries this responsibility is given to the executive. Another significant difference is the power that the parliament has to control the executive. The parliament might legislate to delegate more executive powers to the government, by, for example, allowing it to decide the budget and enforcement policies of the OHS enforcement institutions, such is the case in Britain. Alternatively, the parliament may retain the power to decide the resources and enforcement policies of the OHS enforcement institutions, such as the case in Italy. Thus, these differences are significant in this comparison because the level of criminalisation of OHS crime depends on a few people in Britain, and on the collective decisions taken by all Italian Members of Parliaments.

These criminal justice system and political procedural differences and traditions, therefore, are essential to take into consideration when scrutinising regulatory enforcement policies and institutions. These differences are significant because the enforcement institutions and officers will do anything these procedural powers allow them to do when deciding enforcement activities and while attempting to ensure a higher level of compliance among duty holders. In other words, the legal procedures will affect the decisions taken by the enforcement officers and enforcement institutions, and thus have a significant impact on the enforcement policies they can use, their effectiveness and level of criminalisation of regulatory crimes.

3   Regulatory Styles, Enforcement and Practices

In fact, all these procedural differences become essential to consider when scrutinising regulatory enforcement activities. A fundamental feature of economic regulations is that they have increasingly become principle-based, which means that the objective of the regulation is the achievement of agreed goals, rather than specific OHS standards. Principle-based regulation improves the regulation's adaptability to innovative technologies, which improves safety constantly without governments' or parliaments' need to update standards and regulations. Principle-based regulations are organic statutes capable of adapting to continuous technological advances and changes. However, these types of regulation are not perfect because they are much more interpretable than a standard-based one and, hence, are difficult to enforce either by the regulators or during court trials. This is a fundamental aspect to understand when scrutinising and comparing OHS enforcement policies.

Principle-based OHS regulations have been adopted as the fundamental regulatory philosophy in Europe since the 1989 EU Directive. However, a significant difference is that while the British OHS regulation expects employers to achieve the health and safety of worker as far as it is "reasonably practicable", the Italian one requires the achievement of these goals as far as it is "technologically viable" (Braithwaite 1987; Hutter 1997; Dubini 2001). The difference between these two terms sits quite well within the traditional legal and political differences analysed so far.

According to Edwards v. National Coal Board (1949):

'Reasonably practicable' is a narrower term than 'physically possible' [...] a computation must be made by the owner in which the quantum of risk is placed on one scale and the sacrifice involved in the measures necessary for averting the risk (whether in money, time or trouble) is placed in the other, and that, if it be shown that there is a gross disproportion between them – the risk being insignificant in relation to the sacrifice – the defendants discharge the onus on them. (Edwards v. National Coal Board 1949)

Given this difference, it can be argued that the British regulation allows for the use of cost-benefit analysis (CBA) much more than the Italian one. The evidence-based policies that the Government committed to in the early 2000s aimed at decreasing regulatory burdens through the use of CBA, which in the end only caused a deregulation of the economy and an erosion of OHS standards (Andrews 2007). Reasonably practicable offers flexibility to the enforcement institution and stake holders, it requires them to engage in judgements, supported by evidence when available, while assessing what is a reasonable level of compliance and the costs, including financial, required to achieve it.

In 1997, the European Court of Justice (ECJ) legally challenged the British Government by arguing that the Health and Safety at Work Act 1974 (HASW Act 1974) concept of reasonable practicability was not compatible with the OHS EEC Directive (Great Britain 1974). That is because the reasonable practicability concept implies the permission to undertake CBAs to comply with and enforcing the OHS legislation. In 2007, however, the European Court of Justice rejected the legal challenge and the British Government has been allowed to maintain the concept of reasonable practicability in the OHS law (Tait 2007).

Besides the challenges encountered to create the empirical evidence supporting CBAs, reasonable practicable decisions are also affected by contemporary economic and political climates. In other words, the achievement of compliance also depends on structural social, political, local and global trends and policies (Hutter 1997). This method, however, leads to two further issues. The use of reasonable practicability in the legislation means that the enforcers and duty holders must cooperate in order to achieve compliance (Braithwaite 1987; Hutter 1988). This means that enforcement policies must include a form of cooperation between stake holders and the enforcement institutions, which might lead to conflicts of interest. In addition, the interpretation of *reasonably practicable* might become problematic during court prosecutions. The HSE publishes codes of good practice to help stakeholders understand what the institution means by "reasonable" and "practicable" and how to conduct CBA while enforcing the OHS regulation on premises, which reduces the capacity of legal interpretations (HSE 2015a, b). However, the interpretation of reasonable and practicable in a court of justice might decrease the chances to achieve a guilty verdict. For example, it might shift the burden of proof from the defendant to the claimant, or from the duty holder to the enforcement institution. This issue does not apply for the term *technologically viable* because in this case the requirement for the enforcement institution and duty holders to assess the viability of health and safety preventive measures implemented is reduced, and because the limit of legal responsibility is dictated by the available technology, rather than broader interpretations. Therefore, the use of reasonable practicability can be considered as an enforcement tool offering flexibility to the enforcers and stakeholders, but also a method increasing the ambiguity of the law and posing tangible barriers to the prosecutions of OHS crimes.

These procedural traditions and barriers, and ambiguous legal definitions can, thus, cause a series of reactions from the enforcement institutions, especially if these have the discretionary power to change the enforcement policies they use, as is the case in Britain. In Britain, the regulatory enforcement policies adopted, thus, do not only derive from the best practices ensuring regulatory compliance among stakeholders, but also the most efficient, legally and financially, for the enforcement institution. Hence, regulatory enforcement policies should be scrutinised by their capability of achieving the best level of regulatory compliance in the most efficient way for the enforcement institution, and this is more the case in Britain than in Italy.

Indeed, a fundamental aspect in the analysis of OHS regulatory enforcement activities is the enforcement approaches to use to ensure legal compliance. This topic is, indeed, the central focus of this book. Law enforcement strategies in this field are fundamentally divided into two broad ideas. These are, whether it is better to enforce the law by adopting deterrent techniques based on penalties and prosecutions, or by using assistive techniques based on education and cooperation with the regulated community. The comparison between Britain and Italy becomes a great contribution to this argument because the enforcement policies in these two jurisdictions diverge significantly and understanding the reasons why is valuable.

Conceptually, law compliance can be ensured in the three methods, by deterring duty holders with retributive actions if they fail to comply with the law (the stick approach), by persuasion and education (the carrot approach), or by a mixture of the two. The latter option is the most used because the enforcement institutions are given the flexibility to be responsive to specific issues, but the enforcement policies adopted on the field will be a mix of deterrent persuasive methods. The emphasis given to these two types of policies by the enforcement institutions can impact significantly the level of criminalisation of these crimes and on the redistributive power of the law or regulation enforced. The literature uses a number of terms to describe these two main approaches, but in this book, I have decided to use deterrence-driven enforcement policies those that attempt to achieve compliance mainly through deterrent law enforcement strategies, and compliance-driven those that attempt to achieve compliance and reform duty holders' behaviours mainly through persuasion and education. It is important also to take into account that, on the one hand, compliance-driven policies can also be deterrent, especially if duty holders fail to respond to the enforcement institution's instructions and requests. On the other hand, deterrence-driven enforcement policies are also compliance-driven since duty holders are persuaded to achieve compliance through threats of on-the-spot penalties or court prosecutions. Hence, this book frames the literature debate, analysis and discussions around the deterrence-driven and compliance-driven enforcement policies.

Since the 1970s the literature has become increasingly divided regarding the policies that regulatory enforcement institutions should use to achieve compliance. The innovative policies theorised since the 1970s argue that enforcement activities based on education and cooperation with the regulated community yield more compliance, and therefore are better to reduce harm (Hawkins 1984, 2002; Baldwin 1990; Bardach and Kagan 1982; Braithwaite 1985, 2002). Compliance-driven policies mostly rely on convincing duty holders to comply by providing help, advice and support. These include the use of policies that entrust duty holders with the responsibility to ensure compliance, such as free-of-charge information and advice for businesses, self-regulation, responsive enforcement techniques and reliance on auditing organisations. This may be for a variety of reasons, but the underpinning assumption is that firms are inherently "good" and willing to comply with the law. This assumption has been the case particularly for OHS regulators in Britain. This practice might also reduce the costs of enforcement and compliance, use resources more efficiently, and promote businesses' social responsibility. However, key findings supporting this thesis are based on the sociological micro-socio interactions between enforcement officers and employers during inspections (see Hawkins 2002). Critics have argued that this analytical approach is too narrow because it does not consider wider social structural issues, the roles that the criminal justice system has in terms of wealth redistribution, justice and equality, or the social costs of OHS incidents. In other words, as Pearce argued in 1976, these analyses and theories fail to critically take into account local and global political and economic forces and interests.

The second school of thought supports enforcement policies based on deterrence-driven policies (Geis 1996; Pearce and Tombs 1998; Davis 2004; Tombs and Whyte 2003, 2007, 2009, 2010; Fooks 2008; CCA 2008). Deterrence-driven policies mostly rely on enforcement tools that require employers to pay, financially or other retributive decisions, for the breaches committed. These include regular visits of OHS enforcement officers granted with the power to use of on-the-spot penalties and the use of courts' prosecutions. The main reason underpinning this approach is that firms are thought to be inherently selfish, profit-seeking and careless of employees' OHS. This is based on the assumption that capitalism's sin qua non, which is profit-seeking and the creation of surplus, inevitably leads to the harm and exploitation of the working class (Tombs and Whyte 2003, 2007). This book has been supported by the argument that businesses are rational entities that engage in CBAs when deciding whether to comply with the law. It can also be argued that this system is more concerned with ensuring the principles of equality before the law and avoiding regulatory capture by ensuring workers' compensation and citizens' rights. While the British OHS enforcement institutions use an approach that reassembles the compliance-driven, the Italians are using an approach that is more deterrence-driven. This book argues that the reason for these differences is caused by fundamental political, economic and criminal justice system values and policies. Some of these values, such as the difference between due-process and crime-control criminal justice system and the separation of powers doctrine, were already explored above.

Another key difference, which this book dedicates two whole Chapters to (6 and 7), is the extent to which enforcement institutions and officers can use discretionary practices to enforce the regulation. Discretion is defined by the English Oxford Dictionary (2017) as "the freedom to decide what should be done in a particular situation", which in this context is referred to the decisions of regulatory enforcement officers and institutions in the course of their interactions with duty holders.

Discretion is an embedded feature of the British criminal justice system. British OHS enforcement officers are allowed to use discretion, but they have to follow the Enforcement Management Model (EMM), which regulates their activities to ensure that the enforcement actions are consistent and hence fair (HSE 2013b). In Italy, discretion is not allowed and the Code of Penal Procedure (CPP), which is a national law enforcement code adopted by all law enforcement authorities' officers, prescribes the enforcement officers' behaviour very precisely. The CPP is rooted in the principles of legal consistency and proportionality and the fundamental Italian criminal justice system practice that enforcement officers can only *report* actions that are criminalised by law, which means that they must not take discretionary subjective decisions. OHS enforcement officers in both countries greatly value discretionary practices, but its use during regulatory enforcement activities seems to cause under-criminalisation. When enforcement actions target crimes committed by people or organisations of upper social status, discrimination often seems to be used to decriminalise deviancy. When targeting people or organisations with less social or political power, criminalisation tends to increase (Bronitt and Stenning 2011). Discretion is a traditional fundamental tool embedded in the British criminal justice system, which allows for flexibility in otherwise strict legal rules, but contravenes basic modern principles of legal consistency, fairness and proportionality. The Italian legal system stresses much more the importance of these strict due-process legal principles, and, thus, perceives discretionary practices as a hazard to these values and the legitimacy of the criminal justice system, and of the civil liberties of citizens. In Italy, discretion is generally injected into law enforcement practices through public prosecutors' decisions, which achieves both flexibility and legal consistency. Hence, enforcement institutions' and officers' freedom to use discretion is crucial in this comparative analysis, but also when this practice is compared with other enforcement factors, such as the funding available to the institutions (Bartrip and Fenn 1980).

In fact, another significant difference is the levels of financial resources given to enforcement institutions, but also the political processes by which these are decided. The importance of the level of resources distributed to public institutions enforcing OHS regulation is a neglected subject in the literature proposing innovative compliance-driven enforcement policies. Indeed, the resources given to OHS enforcement institutions in Britain and Italy and the political processes used to achieve those decisions are significantly different. While in Britain there is more emphasis on the costs and benefits of the enforcement policies adopted, in Italy policy development formation does not allow much freedom to consider the monetary efficiency of the policies adopted. In other words, the British Government has significantly more executive powers over the budget that the OHS enforcement institution receives and on the way these funds are spent, than the Italian one.

The level of resources available has had a significant impact since the 1830s, when the British Factory Inspectorate decided to revert to compliance-driven policies due to their incapability to use deterrence-driven ones (Bartrip and Fenn 1980). While in Britain enforcement institutions resource are decided by the government (without direct parliament involvement), in Italy they are agreed in parliament through financial laws[7] (Pelliccia 2008). Another relevant difference in this context is that the British enforcement institution is allowed, in agreement with the Secretary of State for Work and Pensions (SoSWP), to change the enforcement policies it uses, while in Italy these must also be approved by parliament and, must comply with constitutional principles. Constitutional principles are essential because, for example, these guarantee that enforcement institutions undertake a minimum level of proactive inspections per year, which is not the case in Britain. The difference between Britain and Italy is that in the former the resources available determine the enforcement policies the institution and officers can use, while in Italy the resources given to the enforcement institutions must meet a minimum level of enforcement standard as approved by the parliament, which decisions must comply with constitutional

---

[7]Financial Law or Legge Finanziaria.

principles. Hence, British OHS enforcement policies are affected much more by executive decisions taken by the government in power than in Italy. The lack of parliamentary scrutiny in Britain is problematic in this context, but also a long-standing issue in British politics (Pollitt 2004). While in Britain, the way resources are decided increase political governability, but also lead to more abuses of power and policies that are less just and redistributive; in Italy this is the opposite.

OHS enforcement policies are more affected by politics in Britain, while this is less the case in Italy where constitutional principles limit the government's power to change OHS enforcement activities radically and, effectively, change their level of criminalisation. The Italian political institutions' and politicians' obligation to abide by constitutional principles preserves citizens' rights and socio-economic redistributive mechanisms, but this comes at a price. OHS enforcement institutions in Italy seems to be much more inefficient and unresponsive than in Britain. These differences raise fundamental questions on the role of modern nation-states, the redistribution of political power within it, the preservation of citizens' civil rights, and their commitment to promoting and preserving fundamental modern liberal values.

4   View on Abolitionism and OHS Under-Criminalisation

It is important to mention that this book is not strictly concerned with whether the criminal justice system should be retributivist towards employers and company directors. This book concentrates on the inconsistency of the criminal justice system's response to different types of crimes and other forms of social harms, and the level of social injustice this causes, especially to less affluent social classes. In fact, the author is in favour of penal abolitionism and believes that any form of criminalisation and retributive reaction of the criminal justice system does not represent a constructive and appropriate response to deviant behaviour. Yet, the under-criminalisation of OHS and other regulatory crimes committed by social elites demonstrates a disregard for basic modern principles of social justice and equality from the state and criminal justice system institutions. This causes a considerable amount of harm for OHS victims and their families and friends. The book analyses the issue of the under-criminalisation of OHS crimes as a problem causing social inequality and injustice.

5   Book Roadmap

The book is divided into seven more chapters. Chapter 2 provides an overview of the policies and practices of occupational health and safety (OHS) policies and enforcement institutions. It introduces the reader to the historical development of OHS legislation and enforcement policies in Britain and Italy. It critically analyses the reasons of the law enforcement reforms that have been introduced since the 1970s and, in particular, since the early 2000s in Britain and mid-1990s in Italy. The chapter critically analyses the main enforcement organisations' functions, operation and legal powers. It explores the fundamental differences between the British and Italian criminal justice systems in order to understand the reasons why these two jurisdictions adopt different OHS enforcement strategies despite the similarity of their regulatory frameworks.

Chapter 3 represents a courageous attempt to compare the British and Italian occupational health and safety (OHS) incident trends from the 1980s to the mid-2000s. It compares incident trends per active worker and among main economic sectors. This comparative analysis is challenging due to the fundamental difference in the method adopted to record OHS incidents in Britain and Italy, and the changes that have occurred in the methods used to classify incidents, workers' numbers and economic sectors. The chapter also analyses the available data on the enforcement activities conducted by British and Italian OHS enforcement institutions, and also attempts to critically account for the mis- and under-reporting of these crimes and the wrong social and political perceptions that this causes.

Chapter 4 aims is to highlight the methodological issues that were encountered during the fieldwork that have informed the findings of this book. The research fieldwork lasted for two years, during which the British enforcement institution attempted to discourage the author from interviewing occupational health and safety (OHS) enforcement officers. Empirical investigations concentrating on social elites are challenging because academics often deal with people and organisations located in a higher social hierarchy than themselves. Researching up, thus, is challenging because it is extremely difficult to get access, and these studies are usually affected by complicated ethical and legal issues.

Chapter 5 analyses and compares the relationship between the resources available and the enforcement policies used by occupational health and safety (OHS) institutions, and the political mechanisms to distribute these for enforcement activities. This chapter concludes that justice is expensive and that austerity policies leading to the erosion of funds needed for regulatory enforcement activities have caused a decriminalisation of OHS crime and, hence, injustice and inequality. This chapter also raises fundamental questions on whether enforcement institutions should be provided with enough resources to fulfil a constitutional obligation to achieve their statutory goals, such is the case in Italy, or whether these fundamental constitutional principles should not guarantee enforcement institutions' mandates, such is the case Britain.

Chapter 6 explores the relationship between the use of enforcement discretion and the achievement of fundamental modern principles of legal consistency, proportionality and fairness. The use of legal discretion is embedded in every level of the British criminal justice system but forbidden in Italy, especially within executive's directed law enforcement agencies. The use of discretion gives flexibility to officers and agencies, but it can create inconsistent and unproportionate responses of the criminal justice system to different crimes and social classes. In this context, discretion causes the under-criminalisation of occupational health and safety (OHS) crimes because these breaches are not treated the same as other similarly harmful crimes. In other words, regulatory unreasonableness caused by the strict application of the law is not, after all, unreasonable if the reactions to OHS crimes are compared to other similar harmful ones.

Chapter 7 critically analyses how discretionary practices affect the outcomes of occupational health and safety (OHS) regulators' and officers' activities in Britain and Italy, and how these can lead to the under-criminalisation of these crimes. It analyses how discretion affects the enforcement activity planning process. This is important because targeting decisions influence the level of criminalisation of duty holders. It analyses the issues faced by enforcement officers when they are empowered to enforce the law both with deterrence-driven policies and compliance-driven ones. The compliance-driven enforcement traditions of the institutions in both jurisdictions create conflicts between the officers' willingness to achieve compliance and the requirements of the legal system to achieve justice. Finally, it analyses how on-the-spot penalties and enforcement charges consist of an economical and effective strategy to achieve compliance and justice.

The conclusion, Chapter 8, is a summary of the book. It reintroduces its scope by exploring the issue of the under-criminalisation of occupational health and safety (OHS) crimes and the reason it is important to take this issue as the pivotal point for the development of future regulatory enforcement policies. The chapter argues that academics should propose law enforcement policies that are capable of achieving consistent and proportionate responses across the criminal law spectrum, rather than just reasonable responses for OHS duty holders. The chapter pools together law procedural traditions and practices, historical decisions and policy developments and incident and enforcement trends to conclude that the under-criminalisation of these crimes is caused by embedded discriminatory policies and practices in contemporary nation-states' legal and political systems.

References

Andrews, P. (2007). Are market failure analysis and impact assessment useful? In S. Weatherill (Ed.), *Better regulation* (pp. 49–82). Portland: Hart Publishing.

Baldwin, R. (1990, May). Why rules don't work. *The Modern Law Review, 53,* 321–332.

Bardach, E., & Kagan, R. A. (1982). *Going by the book: The problem of regulatory unreasonableness.* Philadelphia: Temple University Press.

Bartrip, P. W. J., & Fenn, P. (1980). The conventionalization of factory crime a re-assessment. *International Journal of the Sociology of Law, 8,* 175–186.

Braithwaite, J. (1985). *To punish or persuade: Enforcement of coal mine safety.* Albany: State University of New York Press.

Braithwaite, J. (1987). Negotiation versus litigation: Industry regulation in Great Britain and the United States. *American Bar Foundation Research Journal, 2,* 559–574.

Braithwaite, J. (2002). Rules and principles: A theory of legal certainty. *Australian Journal of Legal Philosophy, 27,* 47–82.

Bronitt, S., & Stenning, P. (2011) Understanding discretion in modern policing. *Criminal Law Journal, 35*(6), 319–330.

Carson, W. G. (1970a, October). White collar crime and the enforcement of factory legislation. *British Journal of Criminology, 10*(4), 383–398.

Carson, W. G. (1970b). Some sociological aspects of strict liability and the enforcement of factory legislation. *Modern Law Review, 33,* 396.

Carson, W. G. (1979). The conventionalisation of early factory crime. *International Journal of the Sociology of Law, 7,* 37–60.

Centre for Corporate Accountability (CCA). (2008). *Fines against most companies convicted following work-related deaths less than 1/700th of their turnover, new research shows* [Online]. Available from: http://www.corporateaccountability.org.uk/press_releases/2008/mar16sent.htm. Accessed 25 Aug 2014.

Council of the European Union. (1989, July 29). *EEC Council Directive 89/391/EEC.* Introducing measures to encourage improvements in the health and safety of workers at work. *Official Journal, L 183,* 1–8.

Cowling, M. (2011). Can Marxism make sense of crime? *Global Discourse, 2*(2), 59–74.

Damaška, M. R. (1973). Evidentiary barriers to conviction and two models of criminal procedure: A comparative study. *University of Pennsylvania Law Review, 506,* 1972–1973.

Damaška, M. R. (1986). *The faces of justice and state authority: A comparative approach to the legal process.* Yale: Yale University Press.

Davis, C. (2004). *Making companies safe: What works?* Centre for Corporate Accountability [Online]. Available from: http://www.unitetheunion.org/uploaded/documents/Making%20Companies%20Safe%20-%20what%20works%20(CCA-Unite%20paper)11-4856.pdf. Accessed 25 Aug 2014.

Davis, M., Croall, H., & Tyrer, J. (2010). *Criminal justice* (4th ed.). Harlow: Pearson Education Limited.

de Secondat, C.-L., & Baron de La Brède et de Montesquieu. (2001). *The spirit of the laws* (Translated from the French, by D. W. Carrithers & T. Nugent). Kitchener, ON: Batoche Books (Originally printed in 1748).

Dubini, R. (2001). *Articolo 2087 del codice civile. L'obbligo del datore di lavoro di attenersi al principio della massima sicurezza tecnologicamente fattibile.* Sicurezza tecnica, organizzativa e procedural [Online]. Available from: http://www.dbworld.it/file/studi/2087_1329822805.pdf. Accessed 25 Aug 2014.

Edwards v. National Coal Board. (1949). 1 All ER 743.

English Oxford Dictionary. (2017). Discretion. In *The English Oxford dictionary* [Online]. Available from: https://en.oxforddictionaries.com/definition/discretion. Accessed 17 Aug 2017.

Fooks, G. (2008) *The relationship between the levels of fines imposed upon companies convicted of health and safety offences resulting from deaths, and the turnover and gross profits of these companies.* Centre for Corporate Accountability [Online]. Available from: http://www.corporateaccountability.org.uk/dl/manslaughter/reform/ccasentresearchmar08.doc. Accessed 25 Aug 2014.

Geis, G. (1996). Definition in White-Collar Crime Scholarship: Sometimes It Can Matter. In Helmkamp et al. (Eds.), *Definitional Dilemma: Can and Should There be a Universal Definition of White-Collar Crime?* Morgantown, WV: National White-Collar Crime Center Training and Research Institute.

Great Britain. (1974). *Health and Safety at Work etc. Act 1974 (c.37).* London: HMSO.

Hawkins, K. (1984). *Environment and enforcement: Regulation and the social definition of pollution.* Oxford: Clarendon Press.

Hawkins, K. (2002). *Law as last resort: Prosecution decision-making in a regulatory agency.* Oxford: Oxford University Press.

HSE. (2013a). *Health and safety statistics* [Online]. Available from: http://www.hse.gov.uk/statistics/index.htm. Accessed 25 Aug 2014.

HSE. (2013b). *Enforcement management model* [Online]. Available from: http://www.hse.gov.uk/enforce/emm.pdf. Accessed 25 Aug 2014.

HSE. (2015a). *HSE principles for Cost Benefit Analysis (CBA) in support of ALARP decisions* [Online]. Available from: http://www.hse.gov.uk/risk/theory/alarpcba.htm. Accessed 15 June 2015.

HSE. (2015b). *Cost Benefit Analysis (CBA) checklist* [Online]. Available from: http://www.hse.gov.uk/risk/theory/alarpcheck.htm. Accessed 15 June 2015.

Hutter, B. (1988). *The reasonable arm of the law? The law enforcement procedures of environmental health officers.* Oxford: Clarendon Press.

Hutter, B. (1997). *Compliance: Regulation and environment.* Oxford: Clarendon Press.

Jones, S. (2013). *Criminology.* Oxford: Oxford University Press.

Langbein, J. H., Lerner, R. L., & Smith, B. P. (2009). *History of the common law: The development of Anglo-American legal institutions.* New York: Aspen Publishers. ISBN 978-0-7355-6290-5.

Lemert, E. (1967). *Human Deviance, Social Problems, and Social Control.* Prentice-Hall Sociology Series. Englewood Cliffs, NJ: Prentice-Hall.

Lemert, E. (1972). *Human Deviance, Social Problems, and Social Control.* Prentice-Hall Sociology Series, (2nd Ed.). Englewood Cliffs, NJ: Prentice-Hall.

Ma, Y. (2002). Prosecutorial discretion and plea bargaining in the United States, France, Germany, and Italy: A comparative perspective. *International Criminal Justice Review, 12,* 22–52.

Mousourakis, G. (2015). *Roman law and the origins of the civil law tradition.* Cham: Springer International Publishing. ISBN 978-3-319-12267-0; e-ISBN 978-3-319-12268-7; https://doi.org/10.1007/978-3-319-12268-7.

Nelken, D. (2010). *Comparative criminal justice: Making sense of difference.* London (UK), Thousand Oaks (CA), New Delhi (IND) and Singapore: SAGE Publications Ltd.

Ogus, A. (1994). *Regulation: Legal form and economic theory.* Oxford: Clarendon Press.

Packer, H. L. (1968). *The limits of criminal sanction.* Stanford, CA: Stanford University Press.

Pearce, F. (1976). *Crimes of the Powerful. Marxism, crimes and deviance.* London: Pluto Press.

Pearce, F., & Tombs, S. (1990). Ideology, hegemony and empiricism: Compliance theories of regulation. *British Journal of Criminology, 30*(4), 423–443.

Pearce, F., & Tombs, S. (1991). Policing corporate 'Skid Rows': A reply to Keith Hawkins. *British Journal of Criminology, 31*(4, Autumn), 415–426.

Pearce, F., & Tombs, S. (1998). Foucault. *Governmentality, Marxism, Social & Legal Studies, 7*(4), 567–575.

Pelliccia, L. (2008). *Il nuovo testo unico di sicurezza sul lavoro.* Rimini: Maggioli Editore.

Pollitt, C. (2004). *Unbundled government: A critical analysis of the global trend to agencies, quangos and contractualisation.* London: Routledge Studies in Public Management.

Rawls, J. (1973). *A theory of justice.* Oxford: Oxford University Press.

Reiman, J., & Leighton, P. (2010). *The rich get richer and the poor get prison: Ideology class and criminal justice.* Boston: Allyn & Bacon.

Robens, L. (1972). *Report of the committee on safety and health at work 1970–1972.* London: HMSO.

Rogers, J. (2006). Restructuring the exercise of prosecutorial discretion in England. *Oxford Journal of Legal Studies, 26*(4, Winter), 775–803.

Sabine, G. H., & Thorson, T. L. (1973). *A history of political theory.* Hinsdale: Harcourt Brace.

Sanders, A., & Young, R. (2007). *Criminal justice* (3rd ed.). Oxford: Oxford University Press.

Sutherland, E. (1949). *White-Collar Crime.* New York: Holt Rinehart & Winston.

Sung, H.-E. (2006, May). Democracy and criminal justice in cross-national perspective: From crime control to due process. *The ANNALS of the American Academy of Political and Social Science, 605*(1), 311–337.

Tait, N. (2007). ECJ rejects challenge to health and safety laws. *Financial Time,* 15 June 2007. [Online]. Available from: http://www.ft.com/cms/s/df854548-1add-11dc-8bf0-000b5df10621,Authorised=false.html?_i_location=http%3A%2F%2Fwww.ft.com%2Fcms%2Fs%2F0%2Fdf854548-1add-11dc-8bf0-000b5df10621.html%3Fsiteedition%3Duk&siteedition=uk&_i_referer=http%3A%2F%2Fsearch.ft.com%2Fsearch%3FqueryText%3DECJ%2Brejects%2Bchallenge%2Bto%2Bhealth%2Band%2Bsafety%2Blaws#axzz3BRR260SX. Accessed 25 Aug 2014.

Taylor, A. J. (1972). *Laissez-faire and state intervention in nineteenth-century Britain.* London: Palgrave Macmillan.

Tombs, S., & Whyte, D. (Eds.). (2003). *Unmasking the Crimes of the Powerful: Scrutinising States and Corporations.* New York and London: Peter Lang.

Tombs, S., & Whyte, D. (2007). *Safety crimes.* Cullompton: Willan Publishing.

Tombs, S., & Whyte, D. (2009). A deadly consensus: Worker safety and regulatory degradation under New Labour. *British Journal of Criminology, 52*(5), 997–1016.

Tombs, S., & Whyte, D. (2010). *Regulatory surrender: Death, injury and the non-enforcement of law.* Liverpool: Institute of Employment Rights.

Vile, M. J. C. (1963). *Constitutionalism and Separation of Powers.* Oxford: Clarendon Press.

Warwick, P. (2006). *Policy Horizons and Parliamentary Government.* Basingstoke, Hampshire: Palgrave Macmillan.

2

Histories and Traditions

Occupational health and safety (OHS) regulations developed in Europe as a pragmatic response to social problems caused by the industrialisation of economies (Taylor 1972 as cited by Ogus 1994). While Britain implemented the first OHS regulation in the 1820s (Bartrip and Fenn 1980a, b, 1983), Italy only started to regulate OHS at the turn of the twentieth century (Slapper 2000). The post-World War II era witnessed an expansion of the OHS regulation across Europe and a proactive involvement of the state in the enforcement activities (Dawson et al. 1988). The first goal-setting regulation was implemented in Britain in 1974, and this was adopted as the main OHS regulatory philosophy by the European Economic Community (EEC) in 1989. In 1994, Italy complied with the European OHS regulatory philosophy.

Since 1974 in Britain and 1994 in Italy, OHS regulations have witnessed marginal amendments. In the 1990s the European member states' OHS regulations became bound to the 1989 European Union (EU) Directive, which also encouraged jurisdictions to achieve regulatory goals through both traditional deterrence-driven and compliance-driven enforcement policies (European Commission 2007, 2012). Since then, policymakers, especially in Britain, have turned

© The Author(s) 2019
D. Canciani, *The Politics and Practice of Occupational Health
and Safety Law Enforcement*, Critical Criminological Perspectives,
https://doi.org/10.1007/978-3-319-98509-1_2

their attention to the enforcement policies adopted to achieve regulatory goals. However, while in Italy these reforms have been introduced to comply with the EU regulation, in Britain these reforms have mostly been driven by political rhetoric criticising the burden of economic regulation and the need to create an attractive economic environment for businesses and to secure the British economy's competitive advantage in the global market (Posner 2003).

This chapter provides a historical context of the development of the OHS legislation in Britain and Italy, and an overview of the responsibilities, powers and decision-making process of the main institutions responsible for enforcing the OHS regulations. The chapter is divided into two broad parts: the British sections and the Italian sections. The historical analysis for both jurisdictions is mostly focused on the development of the regulation and enforcement policies since the 1970s and, in particular, since the early 2000s in Britain and mid-1990s in Italy. After analysing the historical development, both the British and Italian sections of the chapter will analyse the functioning and organisation of the main enforcement institutions as well as the enforcement tools they are empowered to use. Each section will also analyse the procedures enforcement officers must abide by, the penalties or charges they can issue on the spot and the predicted sentences OHS crimes can lead to at the end of court prosecutions.

1   Britain

The first British OHS legislation and enforcement institutions were formally instituted with the British Factory Act 1844 (Bartrip and Fenn 1980b). In terms of sentences, OHS legislation in Britain developed through the series of Factory Acts that gradually increased workers' OHS rights through the first half of the twentieth century and in the post-WWII period. From the interwar period OHS legal requirements and enforcement institutions gradually increased in volume and by the end of the 1960s there were calls to reform the legislation. This happened because of a series of work-related disasters, which led to the criticisms that the regulatory framework and enforcement policies had become obsolete. Among other critiques, the regulation had reached a level of over-complexity (Dawson et al. 1988). Hence, in 1970 the Conservative government instituted the Robens Commission which was required to investigate the reasons that the Factory legislation failed to meet its aims (Tombs and Whyte 2007; Hill 2006). The *Robens Report* led to the implementation of the Health and Safety at Work Act 1974 (HASW Act 1974), the main OHS legislation in the UK (Dawson et al. 1988; Great Britain 1974).

The *Robens Report* and its recommendations became the foundation of the new HASW Act 1974 (Tombs and Whyte 2007; Woolf 1973). The HASW Act 1974 is based on three principles. Firstly, employers were given the responsibility to manage strategies to decrease incidents in their working environments (Dawson et al. 1988). Secondly, it instituted a form of tripartite compliance system between employers, workers and enforcement institutions: employers are legally liable for employees' health and safety, workers are expected to cooperate in the achievement of health and safety standards in the workplace, and enforcement institutions check that employers and workers cooperate adequately (Dawson et al. 1988; Woolf 1973). Thirdly, it also assigned each employer the duty "to ensure, so far as is reasonably practicable, the health, safety and welfare at work of all his employees" (HASW Act 1974 Section 2(1)). Reasonable practicability meant that the regulation was capable of adapting to different and evolving work environments through a flexible interpretation of the level of standards to achieve in each work situation (Dawson et al. 1988). In the 1949 case of Edwards vs National Coal Board it was decided that reasonable practicability implies a form of cost-benefit analysis (CBA) between the achievement of the suitable standards and the costs, work and time needed to achieve it. Failure to reach reasonable standards of safety should lead to a discontinuation of work activities (Dawson et al. 1988; Rideout 1989).

From the implementation of the HASW Act 1974, and especially during the Conservative government years of the 1980s and 1990s only a few pieces of OHS regulations were introduced. The labour reforms implemented in that period also challenged the implementation of key aspects of the HASW Act 1974 (Dawson et al. 1988). Between the end of the 1970s and the end of the 1980s the OHS regulations implemented were few and mostly reformed the incident reporting systems. By the 1990s the Health and Safety Executive (HSE), the principal British OHS enforcement institution, had the largest numbers of front-line agents in history. In addition, it was observed that the yearly improvement rate of OHS incidents started to level. The election of the Labour government in 1997 led to a revitalisation campaign of OHS regulations and enforcement policies (DETR 2000).

The enforcement policies currently adopted by the HSE have been the result of nine controversial enforcement strategies (Davis 2004; Tombs and Whyte 2010). These can be divided into two main stages. The 2000 *Revitalising Health and Safety* report was an initiative consisting of a consultation of all OHS stakeholders in Britain (DETR 2000). The initiative proposed 44 action points to be achieved to revitalise British OHS. These proposals promoted a combination of compliance-driven and deterrence-driven enforcement policies, such as assigning corporate directors direct responsibility for OHS crimes and improving the information database available to stakeholders (DETR 2000; Davis 2004; The Law Commission 1998). This report showed a tangible commitment to curb the under-criminalisation of OHS crimes with both deterrence-driven and compliance-driven strategies. However, this was short-lived. By 2004 the Health and Safety Commission (HSC) programme to revitalise OHS resulted in a more compliance-driven policy approach, which ignored many action points of the 2000 *Revitalising Health and Safety* report (Davis 2004).

Since 2004, the HSE has undertaken a complete U-turn with their regulatory approach (HSE 2004; Davis 2004). The 2004 *Strategy for workplace health and safety in Britain to 2010 and beyond* downplays the importance of the deterrence-driven enforcement policies that were proposed in 2000. The reports mention that conventional deterrence-driven enforcement policies are an important "compliance motivator for *some* employers" (HSC 2004a, p. 11, emphasis added), rather than for most employers. As a consequence, the HSC decided that it would not pursue any more deterrence-driven enforcement policies and those capable of strengthening worker's rights to compensation (Davis 2004). Shortly after the 2004 report, the House of Commons Work and Pensions Select Committee (2004) urged the HSC and the government to implement more of the deterrence-driven policies suggested in the 2000 report, such as increasing penalties for corporate entities and increasing the number of enforcement officers. Despite these recommendations, compliance-driven enforcement strategies continued to be favoured over deterrence-driven ones.

The 2005 *Reducing administrative burdens: effective inspections and enforcement* report (the "*Hampton Report*") suggests that resources should be "released from unnecessary inspections [and] redirected towards advice to improve compliance" (Hampton 2005, p. 1). This report represented another attack on deterrence-driven policies with a recommendation to use compliance-driven ones based on the provision of advice and information (*Hazards Magazine* 2005; Tombs and Whyte 2010). As a consequence, the HSE demonstrated its commitment to comply with the *Hampton Report*'s deregulation agenda by publishing the 2005 *Sensible health and safety at work: the regulatory methods used in Great Britain* (HSC 2005a). The report, however, suggested that "inspection and enforcement will remain vital intervention strategies [because they are] highly effective in ensuring workplace compliance" (HSC 2005a, pp. 7–8). The report argues that deterrence-driven policies "will remain at the heart of our strategy [even if they] are resource intensive compared to some other methods" (HSC 2005a, p. 12, Section 38). In other words, this report confirmed that deterrence-driven enforcement policies are both effective enforcement strategies, but also expensive to implement. Although this report was the first to appreciate the effectiveness of deterrence-driven policies since 2004, the government's regulatory agenda continued (Tombs and Whyte 2010).

In 2007, the government started planning on weakening the proactive enforcement policies' targeting techniques. The *Regulators' Compliance Code: Statutory Code of Practice for Regulators* proposed a "risk-based proportionate and targeted approach to regulatory inspection and enforcement" (Better Regulation Executive 2007, p. 5). Tombs and Whyte (2010) argue that this is one of the most extreme reports undermining OHS in Great Britain. In fact, this code strongly encourages the use of CBAs when deciding the enforcement policies (Better Regulation Executive 2007). However, it has been demonstrated that CBA used in a regulatory context only represents a justification for deregulating the economy (Posner 2003) because it is used to weigh the measurable monetary costs of the regulation against the unmeasurable social benefits achieved by it (Pearce 1983). This means that in a social policy context, CBA must take into account also moral and ethical social outcomes, rather than only monetary ones, but these cannot be quantified easily, and it is usually preferred over other methods used to measure aggregated choices, such as political referendum (Baldwin and Veljanovski 1984).

The Corporate Manslaughter and Corporate Homicide Act 2007 made corporate entities responsible for legal breaches, but did not make corporate directors and managers responsible for OHS crime. Thus, it did not implement any of the most important *Revitalising Health and Safety* report recommendations. This means that corporate entities can still use this policy weakness to allocate responsibilities to company directors and avoid punishment. Through threats of starting redundancy programmes or by declaring bankruptcy, firms can persuade courts of justice, judges and prosecutors to decrease the severity of penalties and sentences (McBarnet et al. 2007). Therefore, it could be argued that the Corporate Manslaughter and Corporate Homicide Act 2007 improves the chances of making corporate entities more liable for OHS crimes, but their importance in terms of revenues for the state, as well as jobs for workers, means that it is still ineffective in prosecuting these crimes and achieving the purpose of the regulation (Gunningham 2007; Almond 2013; Tombs 2017).

The 2015 HSE Enforcement Policy Statement explains that the goals of the HSE are

[a] ensuring action is taken immediately to deal with serious risks; [b] promoting and maintaining sustained compliance with the law; [c] and ensuring that those who breach the law, including individuals who fail in their responsibilities, may be held to account (this includes bringing alleged offenders before the courts in Britain and Wales, or recommending prosecution to the COPFS in Scotland). (HSE 2015c, p. 2)

The statement clearly shows how they intend to enforce the regulations with compliance-driven policies. This is especially evident in the third statement, which depicts clearly the uncertainty shown by the HSE in "*ensuring* that those who breach the law [...] *may* be held to account" (HSE 2017, p. 2, emphasis added). In other words, despite the lexical incongruence, the statement does not offer certainty of prosecution even if there is enough evidence available.[1]

The last strike to deterrence-driven OHS enforcement policies and the protection of workers' rights came in 2010 when the newly elected Conservative government continued with the deregulation and under-enforcement agenda started by the New Labour government. Following Lord Young's report recommendations (Lord Young of Graffham 2010), the government announced that by 2015 the HSE budget would be reduced by 35% (*Hazards Magazine* 2010). By May 2011, the Department of Work and Pensions report *Good Health and Safety, Good for Everyone* suggested "a very substantial drop in the number of health and safety inspections carried out in the UK" (Department of Work and Pensions 2011a, p. 3). This report also introduced the Fee for Intervention (FFI) policy, which, since October 2012, requires employers to pay a fee when enforcement officers issue suggestion, improvement or prohibition notices (Department of Work and Pensions 2011b; Great Britain 2012). The Löfstedt report also aimed to curb the increasing compensation culture for OHS incidents, even though the amount of compensation for these incidents decreased by 60% since 2003 (Dugan 2013). Hence, the reform was a deliberate attack on workers' rights to compensation (Department for Work and Pensions 2011b). The 2016 *Helping Great Britain Work Well* strategy continues to indicate a minimum emphasis and over-cautious commitment to deterrence-driven policies from the HSE (Tombs 2016).

1.1 Enforcement Authorities

The HASW Act 1974 assigned to the HSE, an executive non-departmental public body, the duty to enforce the OHS regulations. The HSE board can decide the enforcement policies, but these must be agreed

---

[1] The term *ensuring* (guaranteeing to do something) placed in a statement containing a conditional term like *may* (indicating a certain measure of likelihood or possibility) confer contrasting meanings to the statement.

by the Secretary of State for Work and Pensions (SoSWP) (HASW Act 1974). In other words, the SoSWP, which is accountable to Parliament for HSE's activities, has the power to veto the decisions taken by the HSE CEO and Management Board. The system used to decide enforcement policies in Britain is different from the Italian one, where these must be legally approved by Parliament. The SoSWP also has the responsibility to make local authorities (LAs) responsible for the enforcement of relevant statutory provisions. This arrangement is currently coordinated through the HSE/LAs Enforcement Liaison Committee and the HSE/Local Government Panel (HSE 2014). LAs enforcement consistency, proportionality and targeting technique is led by the National Enforcement Code (HSE 2018a). The HSE and LAs enforcement policies are not enacted through statutory means, but through these two committees and the SoSWP (through government agendas), and in agreement with other relevant local, national and international agreements (HSE 2014). The main difference between HSE and LAs' responsibilities is the enforcement allocation. The former is responsible for enforcing the regulation in industries including agriculture, livestock management, dairy production and distribution, construction, factories, mines and quarries and offshore installations. The latter is responsible for enforcing the regulation in the wholesale distribution, warehouse, retail, hotel, catering services, consumer services, leisure industry, and offices (HSE 1998, 2018a).

1.2   Enforcement Activities

The primary function of the HSE control activities is based on the HSE Policy Statement principles, which are proportionality, consistency, targeting, transparency, and accountability (HSE 2015c). These principles drive the decisions enforcement institutions and officers take when conducting proactive and reactive enforcement activities.

1.2.1 Enforcement Targets

In terms of enforcement targeting, since the *Hampton Report* and the *Regulators' Compliance Code: Statutory Code of Practice for Regulators* the

HSE and LAs have started to use a risk-based approach that prioritises high-risk work activities and the most dangerous hazards within them. HSE enforcement activities occur through a bidding system between the Field Operation Directors, HSE Sectors and the Cross-Cutting Intervention Department, which jointly agree on the annual enforcement targets for each HSE field operational sector. Since 2011, these have been affected by the Department for Work and Pensions' commitment to embrace Lord Young's 2010 report suggestion to reduce the funding conveyed to proactive enforcement activities and, hence, the number of inspections conducted (Department of Work and Pensions 2011a; Tombs 2016).

There are three main operational divisions within the HSE. These are the Field Operations Directorate (FOD), the Chemical Industries and Specialist Industries (HID) and the Nuclear Division (ND). Based on the type of activity and the type of risks, the enforcement officer, in agreement with the line manager and other relevant personnel, identifies the type of resources needed for the inspection of a specific site, and makes plans. LAs use similar techniques to select the workplaces to inspect. Despite the specific techniques by which working sites are selected and targeted, the enforcement officers are also required to report other concerns and hazards observed during inspections (HSE 2013a).

1.2.2   Enforcement Tools

The HSE and LA officers' enforcement legal and non-legal tools can be used to achieve the goals set in the Enforcement Policy Statement (HSE 2015c). There are four types of legal tools. The first is offering information and advice to duty holders, in person or in writing, such as suggestions or written warnings. The FFI reform that came into effect in October 2012 has introduced a significant new change to the HSE enforcement policy. For the first time in the history of British OHS enforcement institutions, officers have acquired the power to charge duty holders on the spot following *significant* breaches of the regulation detected during proactive activities. These enforcement practices do not have legal consequences, but might be used as evidence in future legal actions. The FFI policy will be analysed further in Chapters 5 and 7 as it has changed the nature of the OHS regulatory law enforcement. The second is issuing improvement and prohibition notices. The former requires duty holders to fix hazardous work situations within a specific time, while the latter requires duty holders to stop work activities until the hazard has been removed. In both cases, the enforcement officers are required to revisit the workplace. The third is to withdraw approvals and vary licence conditions or issue exemptions. The final legal tool is prosecution. This starts with the issuing of a written legal caution informing duty holders that there is a real prospect that they will be subject to legal prosecution. The decision to continue the prosecution process is undertaken within the HSE or LA (HSE 2015c). Prosecutions will be further analysed in the following section of this chapter.

There are two non-legal tools. The first is Approved Codes of Practice, which are HSE guidelines on how to achieve compliance through approved practices and technologies (HSE 2015a, b, 2018b). While it is not compulsory for employers to follow the Approved Code of Practice, during court prosecutions these are used to shift the burden of proof from the claimant to the defendant (employers), who become responsible for demonstrating that practices, mechanisms and technologies adopted were reasonably adequate to avoid incidents (HSE 2018b). The second non-legal tool is the use of discretionary advice, guidance, and self-assessment aids. Discretion allows enforcement officers to tailor their decisions to particular enforcement situations, but this has been criticised for running counter to the principle of enforcement consistency. For this reason, since 2002, the HSE has adopted the Enforcement Management Model (EMM) which helps the enforcement institution to make enforcement decisions. The EMM is attuned with the Enforcement Policy Statement, the Regulatory Compliance Code and the Legislative and Regulatory Reform Act 2006 (HSE 2013b, 2018b). LAs have also started to use the National Local Authority Enforcement Code, which is similar to the HSE version, but adjusted to enforce OHS in less hazardous workplaces, such as the service sector (HSE 2013c, 2018a).

Despite the HSE commitment to use the EMM, campaigning organisations have criticised the HSE for failing to take firms to court despite available legal evidence, due to a lack of funding (*Hazards Magazine* 2015; UNITE 2008). This issue has also happened despite the work-related death protocol for liaison agreed between the HSE, the Crown Prosecution Service (CPS) and local governments, and other associations, agencies and councils since 1998 (HSE 2016 Work-related Deaths Protocol, January 2016). Hence, although the EMM is used to ensure legal consistency and proportionality and the HSE sets out quite detailed guidelines on the method to select firms for prosecution, these appear, at times, to be ignored. These failures defeat the rationale behind the use of the EMM and compromise the consistency and proportionality in the HSE responses to OHS crimes (*Hazards Magazine* 2015). Therefore, although the HSE's officers have precise enforcement tools and a prescriptive EMM, it has been criticised for failing to take prosecutable firms to court, and a lack of funding seems to be one of the main reasons for it.

The HSE and LAs also have the power to investigate OHS incidents in order to find the causes that led to the incident, to consider each case for prosecution, and improve guidance for stakeholders. This is to achieve regulatory compliance, to fix legal breaches, and to take any necessary enforcement action (HSE 2015c). The HSE Enforcement Policy Statement and the National LA Enforcement Code are based on the principle of legal proportionality, consistency and fairness, but their enforcement actions towards serious OHS breaches has decreased in past year due to funding cuts (HSE 2015c). The HSE yearly business plans set the financial resources devoted to investigations. When the enforcing authority receives complaints and notifications of non-fatal incidents, in order to intervene they consider the severity of harm caused, the gravity of the offence that caused the incident, the employer's historical compliance records, the regulator's enforcement priorities, the likelihood of resolving the issue through investigation, and any public concerns and interests over the event (HSE 2015c). The use of these parameters depends on the resources available and the policy direction adopted by the institution, especially since the early 2010s (Tombs and Whyte 2010).

The HSE's decision to start a prosecution process comprises a number of stages[2] (HSE 2015c). The enforcement officer is the primary decision maker and, where possible, the decision to propose prosecution will be made after consulting the line manager. The officer then presents the case for prosecution to the HSE Approval Officer, which is responsible for ensuring that the proposal document meets the standards expected by HM Inspectors. The standards must follow the Code of Crown Prosecutors, which is a public document issued by the Director of Public Prosecutions. This establishes the general principles that Crown Prosecutors, such as the HSE, should follow when considering laying a case in court (The CPS 2013). For a prosecution to be approved, there must also be sufficient evidence, a realistic prospect of conviction, and it must meet the public interest test (HSE 2015c).

The enforcing authorities should always try to convict individuals as opposed to enterprises, and ensure they publish the names of the companies prosecuted through the HSE website and the media. The OHS enforcement institutions can also take an active role in suggesting and guiding courts on the best sentence to issue, or whether the case should be taken to the Crown Court (HSE 2015c). The reason the HSE seeks to raise penalty levels is to increase the deterrence effects of the enforcement actions and, hence, of the resources spent when using deterrence-driven enforcement policies. It also improves the efficiency of the resources allocated for prosecutions, which represents one of the Hampton principles (HSE 2015c).

The 2005 *Hampton report* has proposed the use of financially efficient enforcement policies by using targeting techniques (Hampton 2005). As a result of the Hampton review, and from the evidence collected by the HSE (Davis 2004; HSE 2003), deterrence-driven enforcement policies are the best known to ensure compliance. However, with the general leniency of penalties issued after sentences,[3] the use of prosecutions

---

[2]In Scotland the prosecution process is different than in England. It is important to note that these models are different in principle and that a lengthier and more in-depth analysis of these two jurisdictions' criminal justice system policies might offer a richer account of their differences. Unfortunately, a lengthier and deeper analysis of the two jurisdictions' criminal justice systems is beyond the primary scope of this research study.

[3]The new 2013 sentencing guideline has increased penalties.

might be the riskiest enforcement tool adopted, which might also explain why the HSE prefers compliance-driven policies over deterrence ones (Hawkins 2002). In addition, since 2008, when investigating a work-related death, the enforcing institutions have to consider whether the dynamics in which the case occurred might justify a manslaughter case. In those cases, the enforcing authorities report the case to the CPS, and the police take over the investigation and prosecution (HSE 2015c). Finally, Crown bodies are not subject to statutory enforcement such as prosecution, but they can still be issued non-statutory improvement and prohibition notices (HSE 2015c).

1.3   Penalties and Sentences

Legal penalties in Britain can only result from court sentences, which means that duty holders can only be financially penalised after prosecutions (HSE 2012b). The FFI policy, however, has also introduced an on-the-spot penalising tool, but this is a charge rather than a judicially imposed penalty (HSE 2012a). Maximum penalties for OHS breaches have increased since the HASW Act 1974 came into force. At purchasing power parity in 1974, duty holders could be penalised a 2013 equivalent of £5400 for most of the offences in lower courts, or receive a maximum imprisonment term of two years or an unlimited fine or both in high courts. In 2008, the Health and Safety (Offence) Act has increased the maximum penalty issuable by lower courts to £20,000 and a twelve months' imprisonment sentence for lower courts, or unlimited penalties and/or two years' imprisonment sentence (Health and Safety (Offence) Act 2008). These penalties are higher than those used in Italy, but until the recent introduction of the FFI policy, these were only issued after prosecutions and therefore quite rare when compared to those issued in Italy. It has also been demonstrated by Fooks (2008) that public limited companies found guilty of causing OHS-related fatal incidents pay penalties amounting to only an average of 0.14% of their turnover. It is interesting to note that already one of the recommendations of the 1972 Robens Committee was to increase punishments through higher penalties and tougher sentences. Although penalties and prosecution sentences issued to non-compliers have increased since the HASW Act 1974, even the HSE "believes that the current general level of penalties does not properly reflect the seriousness of health and safety offences" (HSE 2013a). One of the reasons for this is because prosecuting firms can be a very risky enforcement strategy when the chance of winning is low and resources scarce. The effect of the 2013 new sentencing guideline should increase the level of criminalisation of OHS crimes, but this is yet to be assessed.

2   Italy

The Italian OHS enforcement policies are different from the British despite the similarities between the two regulatory frameworks. Before exploring the Italian OHS regulatory context, however, it is important to understand and reiterate some fundamental features of the Italian legal systems and traditions.

As was mentioned in the introduction, Italian law is based on Roman law, a codified tradition and driven by due-process legal principles, which stress the importance of the processes by which the criminal justice system operates, rather than its outcomes, such is the case in the British crime-control legal tradition (Mousourakis 2015). This means that in order to ensure citizens' civil liberties, the relationship between the state and its responsibilities, and citizens and their duties, is much more regulated than in Britain. In Italy, law enforcement agencies' and officers' behaviours are regulated by the Penal and Civil Codes (CP and CC), the Codes of Penal and Civil Procedure (CPP and CPC) and the enforcement officers' Judiciary Police Official (UPG) law (Repubblica Italiana 1930, 1942, 1988, 2012; Delle Fave 2013). Italian law enforcement officers with UPG powers do not have discretion and can only operate at arms-length from public prosecutors working for the Public Ministry (PM). Public prosecutors have a certain level of discretion to instruct officers on specific enforcement activities. Unlike in Britain, when evidence of criminal conduct is reported to the PM, public prosecutors have to comply with the legal principle of legal obligation, which means that they cannot decide to interrupt a legal prosecution if there is sufficient evidence of criminal conduct (Delle Fave 2013; Fioravanti 2011; Council of Europe 2010). This is very different in Britain (and in England and Wales in particular) where enforcement institutions or the CPS have no strict obligation to continue prosecution processes. These key aspects highlight the difference between the due-process and crime-control legal systems when attempting to achieve legal proportionality and consistency. These principles have great significance for the operations of OHS enforcement officers. This account of the Italian legal system is essential to understand the following historical development of the OHS regulatory enforcement policies in Italy and the legal limits within which officers must operate.

The remaining part of this section analyses the historical development of the Italian OHS legislation since 1994. The second section introduces and explores the Italian OHS enforcement institutions and their duties. The third section analyses the enforcement activities and policies adopted, how the annual enforcement targets are decided and how investigations are conducted. The last part explores the OHS penalties issuable in Italy.

Italian OHS regulations were first enacted in 1899 and the labour inspectorate was instituted in 1906. The regulations and the enforcement institutions increased in size during the twentieth century and were reformed in the post-WWII years. The current OHS legislation is Legislative Decree 81 2008 (Repubblica Italiana 2008). This derives from the Legislative Decree 626 1994 (Repubblica Italiana 1994a), which came into force to comply with the 1989 EEC Council Directive on health and safety (89/391/EEC) and introduced a goal-setting philosophy based on the concept of *technological viability* (Pais 2008; Pelliccia 2008; Dubini 2001).

The Labour Inspectorates, which operate under the direction of the Ministry of Labour, were the main OHS enforcement institutions until 1980 (Pais 2008; Pelliccia 2008). Enforcement activities were conducted by inspectors from the OHS enforcement institutions of the Regional Labour Directorate (DRL) and Labour Territorial Inspectorates (ITL). Their activities are still regulated by President of the Republic Decree 520 1955 (Repubblica Italiana 1955). Another important feature of the Italian OHS regulation is the National

Institute for Insurance against Working Accidents (INAIL), which is responsible for enforcing a national compulsory OHS insurance scheme, recording statistics and promoting a healthy and safe work culture through compliance-driven strategies (INAIL 2014a).

In January 1980, the Italian National Health Service came into force with Law 833 1978 and instituted the Prevention Services for Safety in Work Environments (SPSAL) the main contemporary Italian OHS enforcement institution[4] (Repubblica Italiana 1978; Rubini 2011; Porreca 2008). ITLs retained the responsibility to enforce the OHS regulations in only a few hazardous sectors of the economy (Rubini 2011; Porreca 2008). Between 1980 and 1994, SPSAL enforcement responsibilities were subject to more discretionary powers than ITL ones. This important difference generated a considerable effect on SPSAL OHS enforcement culture and greatly enriches this comparative analysis.

Law 833 1978 heralded, for the first time, the introduction in Italy of compliance-driven enforcement policies, which empowered SPSAL officers to promote good OHS practices among duty holders. The only deterrence enforcement actions that SPSAL officers could use from 1980 to 1994 were improvement and prohibition notices, but these, similarly to the practices in Britain before the 2012 FFI reform, did not lead to any on-the-spot penalty (or charge), and the decision to issue them was not strictly defined by law. In other words, between 1980 and 1994, SPSAL officers did not have UPG powers and obligations. Meanwhile, the 1981 depenalisation law (Repubblica Italiana, 1981) reformed the CP and CPP and, effectively, decriminalised a large number of crimes in order to decrease the procedural workload of the criminal justice system. The reform, however, introduced mandatory administrative on-the-spot penalties to, arguably, compensate for the deterrence-driven vacuum that the reform created in the legal system (Rinaldi 2012). This meant that the SPSAL enforcement practices became incongruent with the 1981 depenalisation reform. The reform brought by the depenalisation law did not apply to SPSAL officers until the mid-1990s, when Legislative Decree 758 1994 (Repubblica Italiana

---

[4]Please note that this name can change among regions.

1994b), for the first time, conferred on them UPG powers. Hence, in 1994, SPSALs officers started to comply with the UPG law and follow the CPP, and they were stripped of the discretionary powers they had enjoyed since 1980.

The reason for this reform is that between 1981 and 1994, judges and lawyers complained that compliance-driven policies and subjective discretion among SPSAL officers created an unconstitutional clash of interests between the state executive and the judiciary (Rausei 2006). That is because SPSAL enforcement officers were not given UPG powers and duties. Thus, their actions were not proportionate and consistent when compared to other law enforcement institutions, and were, therefore, breaching fundamental constitutional principles (Rausei 2006). In other words, the SPSAL enforcement policies were unfairly decriminalising OHS crimes. This is important because the way the enforcement activities were organised in Italy between 1981 and 1994 were very similar to the current British methods, but this changed due to the inconsistency of the criminal justice system's response to OHS crimes compared to other legal breaches. The enforcement policies adopted under-criminalised OHS incidents and the reasons behind the 1994 enforcement policies reforms can contribute to this comparative analysis and the anglophone literature because the criticism of the SPSAL enforcement processes drew from lawyers before 1994 are very similar to those concerning the HSE today.

Legislative Decree 758 1994 (Repubblica Italiana 1994b) prohibited the discretionary powers enjoyed by SPSAL officers since 1980 and introduced *compulsory* improvement and prohibition notices and administrative on-the-spot judicial penalties. The use of penalties meant that duty holders started to be motivated to comply with OHS regulation *before* enforcement officers' visits (Regione Marche, Giunta Regionale Servizio Sanità, Unità O.O. Sanità Pubblica e Prevenzione 1995; Porreca 2008). Hence, it can be argued that the discretionary practices enjoyed by SPSAL officers until 1994 were eliminated and transferred to public prosecutors within the PM, which supervises Italian UPGs at arms-length. These institutionalised or subjective discretionary policies, however, are highly regulated by law. While British officers' discretion is coordinated by the EMM, which are procedures deriving from HSE, CPS and DWP enacted Enforcement Policy Statement, the institutional discretion adopted by Italian officers is legalised through Acts of Parliament (HSE 2013b, Italian CPP). The result is that since 1994, SPSAL enforcement officers and their institutions have lost discretionary control over the enforcement decisions adopted during OHS enforcement activities (Porreca 2008).

Legislative Decree 626 1994 also reorganised the institutional responsibilities of SPSALs, ITL, and INAIL (Repubblica Italiana 1994a; Rausei 2006; Porecca 2008). Initially, there was an attempt to coordinate the deterrence- and compliance-driven enforcement institutions' activities, but from 2000, responsibilities started to be divided between the SPSAL, the ITL and the INAIL. The new arrangement is that SPSAL remains the main OHS deterrence-driven enforcement institution and ITL has been given the responsibility to enforce mainly contractual labour regulations.[5] ITLs still retain some responsibility to enforce OHS regulations, but these are marginal when compared to the SPSAL activities. Legislative Decree 38 2000 (Repubblica Italiana 2000) assigned INAIL the responsibility of collecting statistical information, enforcing the OHS insurance scheme, recording statistics and improving the culture of OHS at work with compliance-driven policies. INAIL gives grants for businesses wishing to invest and improve their OHS standards (Rausei 2006; Porecca 2008; Rinaldi 2012). This institutional arrangement, which is explored in-depth in the next section, was formalised with the Unified Text on Health and Safety in Work Environments or Legislative Decree 81 2008 (Repubblica Italiana 2008). The latest reforms came into force with Law 122 2010 (Repubblica Italiana 2010), which effectively gave INAIL the responsibility to promote, through compliance-driven activities, OHS in the economy (INAIL 2014a). Between 1994 and 2008, SPSAL enforcement values were gradually reformed (5th Italian interviewee; Bonomi and Marinaro 2009; INAIL 2014a). While SPSAL officers' main assumption was and still is to fulfil their preventive duties by alternating education and information with deterrence-driven enforcement policies,

---

[5]These include labour contracts between workers and employers.

the function imposed by the judiciary authority as dictated by the UPG law, CPP and the Italian Constitution is to enforce the law by using deterrence-driven policies such as penalties, notices and prosecutions. While Italian officers perceive the use of deterrence-driven policies as a method to prevents incidents and, hence, achieve compliance more effectively, in Britain deterrence-driven policies are mostly perceived as a form of unnecessary punishment. British officers argued that inducing duty holders' behavioural change with deterrence-driven policies might have only increased antagonism towards the HSE and the OHS regulation, but this did not happen in Italy where the SPSALs officers perceived that the deterrence-driven policies introduced in 1994 cemented their authority, legitimacy and institutional mandate.

A final issue that was identified in the OHS legislation in the 1990s was on corporate sentencing. Until 2001, the Italian legal system was not designed to charge public limited companies of corporate homicide and manslaughter, because legally speaking corporate entities have the same legal valence of one human being[6] (Fooks et al. 2007). In order to comply with the OECD Convention on Combating Bribery of Foreign Public Officials in International Business Transactions 1997, the Italian government passed Legislative Decree 231 2001 (Repubblica Italiana 2001), which allows courts to penalise firms rather than individuals. This Legislative Decree makes companies accountable for crimes, and the amendments enacted by Legislative Decree 81 2008 (Repubblica Italiana 2008) also include OHS violations (Fooks et al. 2007).

2.1   Enforcing Authorities

The Italian labour services were instituted at the end of the nineteenth century, but the contemporary institutional organisation is routed into the administrations created during the first half of the twentieth century as well as the reforms occurred in the 1980s when national health provisions were consolidated under the newly funded National Health

---

[6]The Italian Constitution (Art. 27) does not consents corporate entities to be tried in a court of justice.

Service (Repubblica Italiana 1978). The Italian OHS enforcing institutions are divided between the insurance administration and two enforcement authorities (Pais 2008). While the former enforces mainly insurance policies and schemes for workers with compliance-driven activities, the latter two enforce OHS legislation with deterrence-driven ones. Hence, as mentioned above, the main Italian OHS institutions are the SPSAL, ITL and INAIL[7] (Rinaldi 2012). In 2015 Legislative Decree 149, 2015 (Repubblica Italiana 2015) formed the National Labour Inspectorate, which main task consists into coordinating and rationalising the enforcement activities between all enforcement institutions and according to the central, inter-regional and territorial commissions for the coordination of the enforcement activities.[8]

SPSAL and ITL are the main OHS enforcement authorities in Italy. The implementation of Law 833 1978, which instituted the contemporary Italian National Health Service, shifted OHS enforcement responsibilities from the Ministry of Labour and Social Policies to the Ministry of Health (Repubblica Italiana 1978). This division of responsibilities has also allowed to record and monitor occupational health sicknesses and diseases. The reform, however, has also created some confusions in terms of the enforcement responsibilities and hierarchical authority between ITLs and SPSALs. This reform heralded the beginning of a period of OHS under-enforcement in Italy, from the early 1980s to the mid-1990s. ITL, an institution subordinate to the Ministry of Labour and Social Policies, was the main OHS enforcement institutions until the 1980s and has since mainly retained the responsibility for enforcing labour regulations, such as workers' labour contracts. ITL also has the sole responsibility for enforcing a few OHS regulations, such as work sites using ionising machinery, and some with SPSALs,

---

[7]The Finance Police and military police Carabinieri are also legally empowered to enforce the health and safety regulation, but they are not technically prepared to do so, and their power is used to support and assist the operations of the SPSALs and DPLs health and safety enforcement officers.

[8]In Italy, these are called: Ispettorato Nazionale del Lavoro; Commissione di Coordinamento Centrale (Legislative Decree 124 2004); Commissione di Coordinamento Interregionale (Prime Minister Decree 14 February 2014); Direzioni Territoriali del Lavoro (Prime Minister Decree 121 2014 and Ministerial Decree 4 November 2014).

such as construction sites. Hence, since 1980 most of the responsibilities to enforce OHS regulations were transferred to the SPSAL. While ITL enforcement officers have always had UPG powers, SPSAL officers acquired these only in 1994 (Repubblica Italiana 1978, 2008; Rinaldi 2012).

The SPSAL is a department of the Local Health Agencies (ASLs) delivering health services at borough level (i.e. medical services for citizens), and hierarchically directed by the Ministry of Health and the Regional Health Service. They are responsible for enforcing and investigating OHS and industrial hygiene legislation in every sector of the economy excluding those enforced by ITL, but including construction sites. Since 1994 they acquired UPG powers, which consent them to issue on-the-spot penalties, but forbade them to use compliance-driven enforcement policies during proactive activities. Compliance-driven activities can still be conducted, but these must be organised through trade unions or trade associations. SPSALs' officers are responsible for conducting medical inspections, stake holders can ask them to visit construction sites to suggest changes related to OHS practices, and offer a free-of-charge open-door service to stake holders (Regione Sicilia, Azienda Sanitaria Provinciale Trapani 2013; Regione Friuli-Venezia Giulia, Azienda per i servizi Sanitari n.1 Triestina 2013a, b; Regione Veneto, Azienda ULSS 13 Mirano 2013).

INAIL was funded as a compulsory national OHS insurance scheme to compensate workers for work-related injuries and ill health. It has five main functions. The first function of INAIL is that it compensates most subordinate workers for incidents. It pays workers' wages on behalf of employers during the convalescence period such as the HM Revenue and Customs does for maternity leaves in Britain (Cataldi 1983). The second function of INAIL is to collect statistics. The system of data collection, however, seems to be much more accurate than the one used by the HSE in Britain. Its compulsory insurance scheme reduces significantly the level of mis- and under-reporting of incidents because it incentivises workers and employers to report them. The third function of INAIL is to conduct, if needed, legal litigations against employers on behalf of injured workers. This role allows to avoid the conflict of interests that British workers' face when thinking of suing their employers to recover the costs of OHS incidents. This service allows to fulfils the redistributive regulatory role better and reduces the economic costs of incidents for workers (INAIL 2014a). The fourth function of INAIL is its responsibility to promote OHS through compliance-driven policies. Since 2000, INAIL acquired the responsibility to promote OHS in the economy and, in particular, help SMEs with funding grants. This reform was the first that clearly divided INAIL's role to promote health and safety through compliance-driven policies from the SPSAL's and ITL's deterrence-driven enforcement roles. This reorganisation has reduced the clashes of interests created between the simultaneous use of deterrence- and compliance-driven policies, which was a central feature of the SPSAL's enforcement activities between 1980 and 1994 (Repubblica Italiana 2008, Art. 11).[9] The last function of INAIL is to detect firms employing uninsured workers. However, its enforcement officers do not have UPG powers, which means that they can only issue administrative penalties aiming at the collection of insurance payments and report firms for prosecution (Cataldi 1983). The reason INAIL is important for this comparative analysis is because its function allows to improve the redistributive effect of the OHS regulation, measure the social and economic costs of regulatory failures accurately, and reduce conflicts of interest between workers and employers and of SPSALs and ITLs officers during their enforcement activities.

2.2   Enforcement Activities

There are various tools that enforcement officers can use to ensure duty holders' compliance. This section explores the enforcement *targeting*

---

[9]It is important to note that Italian workers can receive compensation for OHS from INAIL, which also cover for 100% of the workers' wage. Workers' wages are also paid by the National Institute for Social Security (INPS) to workers who are absent from work for causes not related to health and safety, such as sick leave. However, INPS provides only 80% of the workers' wage during sickness, which means that workers might be reporting some injuries happened while not at work as health and safety incidents in order to receive 100% of the wage while on convalescence. Hence, Italian health and safety incidents might be even over reported, especially for those injuries that can be hidden until returning to work premises.

policies, the *proactive* (pre-planned) enforcement tools, the *reactive* (post-incidents) enforcement activities and the *compliance-driven* activities conducted by SPSAL's and INAIL's officers.

2.2.1 Enforcement Targets

The Italian OHS enforcement *targets* are decided every three years by the Permanent Conference between States and Regions and the Autonomous provinces of Trento and Bolzano,[10] and policies are decided with a risk- and random-based approach. The regions act as a unified organisation under the Regional Comities Coordination[11] (Repubblica Italiana 2007; Conferenza delle Regioni e delle Provincie Autonome 2011). The annual enforcement decisions between state and Regional Committees are published in the Health Pact (Patto per la Salute), which consists of an agreement that regions and provinces use as a guide for deciding targets and fund allocations on specific OHS issues (Rinaldi 2012).

The state can propose to increase enforcement activities conducted by SPSAL, but cannot impose annual proactive enforcement targets (Repubblica Italiana 1997, Art. 4). Currently, the Health Pact sets an enforcement target of at least 5% of firms on the territory. Each region breaks the annual target down among manufacturing, construction and the service sectors of the economy, and in accordance with current regional and provincial OHS incident trends (Rinaldi 2012). Regional enforcing authorities can also organise enforcement actions called 'blitz'. These involve the participation of multiple enforcement institutions, target specific industries or specific OHS issues for a limited period (Fooks et al. 2007).

---

[10]Permanent Conference between States and Regions and the Autonomous provinces of Trento and Bolzano (Conferenza Permanente per i Rapporti tra lo Stato le Regioni e le Province Autonome di Trento e Bolzano) The decisions are published in the Ministry of Health, Regions and Autonomous Provences of Trento e Bolzano Agreement's Protocol (Protocollo d'intesa Ministero della Salute, Regioni e Provincie Autonome di Trento e Bolzano).

[11]As known as Comitato di Coordinamento Regionale.

The decisions to commit to specific enforcement targets (currently at 5% of registered firms) and adopt specific regional enforcement priorities is decided in a hierarchical format by the Central Coordination Commission, first, and the Interregional Coordination Commission, second, and the ITL, third[12] (Legislative Decree 124 2004; Prime Minister Decree 14 February 2014; Prime Minister Decree 121 2014 and Ministerial Decree 4 November 2014). Some regions struggle to meet the 5% annual enforcement target, which causes enforcement inconsistency. Since 2007, the regional and provincial enforcement priorities have been decided through the consultation of the Information Flows,[13] which represents a statistical database created through cross-references of quantitative information collected by Insurance administrations (i.e. INAIL, INPS), Regions, Provinces, the Italian National Health Service and the Italian Chamber of Commerce (Fooks et al. 2007; Tiraboschi and Fantini 2009). ITL chief enforcement officers have an additional discretionary power to choose the specific firms to visit in order to meet the 5% enforcement target and are also able to set permanent enforcement units in firms employing large numbers of workers (9th Italian Interviewee).

2.2.2 Enforcement Tools

SPSALs enforcement officers have a number of enforcement tools to use while inspecting duty holders. The enforcement tools used by SPSAL officers during proactive enforcement activities are deterrence-driven policies and can broadly be divided between those tools used to deal with non-criminal and criminal violations. Non-criminal violations are breaches that do not cause an immediate hazard and only lead to on-the-spot administrative penalties. Criminal violations are those causing

---

[12]As known as the Commissione di Coordinamento Centrale (Legislative Decree 124 2004); Commissione di Coordinamento Interregionale (Prime Minister Decree 14 February 2014); Direzioni Territoriali del Lavoro (Prime Minister Decree 121 2014 and Ministerial Decree 4 November 2014).

[13]As known as the Flussi Informativi.

imminent harmful hazards to people, which can result in penalties, a prison sentence convertible into a penalty or a prison sentence[14] (Rinaldi 2012).

According to the CC and Health and Safety regulation, Italian enforcement officers can take three main enforcement decisions when detecting regulatory breaches (Rinaldi 2012). Firstly, enforcement officers can decide to issue on-the-spot administrative penalties, which can be used for less hazardous breaches posing no immediate harmful risk. The enforcement officers are required to explain what the regulatory breach is and suggest solutions to fix the issue (Rinaldi 2012). Since September 2008 the Sacconi's Directive (Ministero Del Lavoro, Della Salute Delle Politiche Sociali 2008) has defined the minimum threshold by which enforcement officers can start to penalise firms with the term 'substantial'. Enforcement officers have been instructed to ignore formal regulatory breaches that do not have the potential to cause health or safety hazards, such as minor clerical errors.

Secondly, enforcement officers can issue compulsory improvement notices and the suspension of business activities through revocation of licences.[15] These notices can be issued for all labour related legal breaches that are not punishable by prison sentences, judicial penalties[16] or both (Rinaldi 2012; Legislative Decree 124 2004; Repubblica Italiana 2008; Legislative Decree 8 2016). Employers must comply with the improvement and prohibition notices within six months. Compulsory improvement notices are compulsory because enforcement officers are obliged to issue them when specific legal breaches are detected. These enforcement tools consist of a form of institutionalised discretion, which have been available since 1955, but their use has increased since the 1981 depenalisation law was introduced. Improvement and prohibition

---

[14]Administrative penalties were introduced into the Italian sanction system by the "de-penalisation" Law 689 1981. Before 1981 all breaches were criminal. Law 689 1981 decriminalised almost all legal breaches that do not pose an immediate hazard and changed the penal procedures of previous criminal breaches and contraventions into administrative penalties (misdemeanour act).

[15]Known in Italy as Prescrizione Obbligatoria (improvements notices) and Diffida (prohibition notices).

[16]On-the-spot penalties are known as ammende.

notices have decriminalised a series of deviant actions considered recti-fiable, but across all legal breaches, not only OHS ones (Venturi 2014).

These two enforcement tools have given duty holders the oppor-tunity to transform penal breaches into administrative ones if the breach is rectified within a certain amount of time (Repubblica Italiana 1994b; Legislative Decree 124 2004; Repubblica Italiana 2008 art 14 C2; Rinaldi 2012). Only regulatory breaches which leads to prison sentences cannot be decriminalised with improve-ment or prohibition notices.[17] Therefore, when breaches can be dealt with an improvement or prohibition notice, the enforcement officer must make a formal request to the public prosecutor, who judges whether to authorise the enforcement decision (Rinaldi 2012). Once duty holders comply with improvement and prohibition notices the criminal breach is penalised by the minimum set by law or by one-fourth of the maximum issuable. Not complying to notices can lead to jail sentences of three to six months or to penalties between €2500 and €6400 for every breach committed (Ministero del Lavoro e delle Politiche Sociali 2004; Porreca 2008; Rausei 2006; European Commission 2005; Rinaldi 2012).

As for the British officers, the advantages of issuing improvement and prohibition notices consist of it efficiency to achieve compliance quickly and at low cost, by requiring employers to improve criminal viola-tions, changing the nature of the breach from criminal to non-criminal, reducing the penalties received, and decreasing the judicial system workload (Porreca 2008; Rinaldi 2012). The Italian improvement notices should be considered as an institutionally discretionary tool, rather than a subjective one, its use is compulsory when violations fall within specific parameters, and because only public prosecutors can approve enforcement officers' decisions. In addition, these enforcement policies are regulated through law and legislative decrees approved in parliament, and therefore cannot be easily changed to suit enforcing

---

[17]These are firms handling dangerous chemical materials, defined as thermoelectric plans, adopt-ing nuclear technology, producing explosives, those employing more than two-hundred employ-ees, and operating in the extractive industry with more than 50 employees.

institutions' short terms needs (Porreca 2008; European Commission 2005; Rinaldi 2012).

Thirdly, court prosecution is another tool that enforcement officers can use to achieve compliance (Rinaldi 2012; Repubblica Italiana 2008). Prosecution is only used when firms classified as high-risk[18] are found without OHS risk assessments, or if they have not complied with improvement and prohibition notices (Rinaldi 2012; Repubblica Italiana 2008). Improvement and prohibition notices are complied with most of the time (1st Italian interviewee), or about 94% of times between 2001 and 2004 (Fooks et al. 2007). During an incident investigation, Italian enforcement officers must investigate the incident, collect evidence, identifying the duty holders legally liable for the breach, and present the case to the public prosecutor.

SPSAL and ITL officers are also given *reactive* enforcement responsibilities, which means that they are given the task to investigating incidents by collecting legal evidence needed by the public prosecutor to decide whether the case should be prosecuted (CPP art. 55 to 59). The investigation process is activated by (a) the emergency services such as the police, Carabinieri or medics (CP art. 365), or (b) automatically for any OHS related sickness that has caused absence from work for more than 40 days. In both cases the Penal Code instructs UPG to prioritise incidents that have caused or are likely to lead to the worker absence for more than 40 days, that have caused an impairment of an organ or a sense, the loss of a sense, the loss of the use of an organ, an illness that is probably or certainly incurable, the loss of a limb or its function, a permanent and serious speech impairment, or a deformity or permanent disfigurement of the face. Therefore, OHS enforcement officers intervene to collect the evidence needed for prosecution only if the incident has caused serious or very serious injuries to the victim(s), but also when a serious incident is narrowly avoided (see Art. 583 of CP; Rinaldi 2012).

---

[18]Firms handling dangerous chemical materials, defined as thermoelectric plans, adopting nuclear technology, producing explosives, with more than two-hundred employees, and operating in the extractive industry with more than 50 employees.

While in Italy the choice of prosecuting OHS breaches depends on whether public prosecutors consider the amount of evidence available adequate, in Britain (and England in particular) the decision to prosecute also depends on other aspects, such as the prospect of reaching a guilty sentence and the public interest test. In addition, while in Britain the enforcement officers are responsible for taking cases to court, in Italy they can only appear in court as witnesses. This procedure can have massive effects on the enforcement decisions, because the Italian enforcement institution does not bear the prosecution costs and workload of their British colleagues.

2.2.3 Compliance-Driven Enforcement Tools

The Italian public institutions responsible for curbing OHS incidents also adopt compliance-driven enforcement methods. Historically, these activities have been conducted by SPSAL officers, but since the post-1994 reforms and in particular since 2000 these have gradually been given also to INAIL. It might be argued that the devolution of these responsibilities to INAIL represents the political intention to separate the institutions conducting deterrence-driven enforcement policies from those conducting compliance-driven ones, and avoid the conflicts of interest caused when these practices are conducted simultaneously during proactive enforcement activities.

From respectively 1994 and 2000, SPSAL and INAIL were officially given the responsibility to promote, educate and train stake holders in small, medium and micro enterprises (Rinaldi 2012). Since Legislative Decree 626 1994 and Legislative Decree 758 1994 (Repubblica Italiana 1994a, b), policy makers have transferred the power to use compliance-driven policies from SPSALs to INAIL. SPSAL personnel is allowed to promote OHS and provide training, but they are to use them when conducting proactive inspections, hence when officers act as UPG. They have been required to establish specific times to set up projects educating and informing stake holders on specific OHS issues, and INAIL has provided grants designed to help firms' achieving compliance, such as to provide workers' training (Rausei 2008). SPSAL educational projects include seminars and conferences organised in partnership with other institutions, and educational courses for workers and employees, but these activities cannot raise revenues, can be organised only through businesses associations and trade unions (3rd Italian interviewee), and the material produced must be made freely available online. SPSAL offices are also allowed to provide information from their offices (Regione Friuli-Venezia Giulia and Azienda per i servizi Sanitari n.1 Triestina 2013b). In the 2000s INAIL made available over €204 million in funding grants for SMEs, by 2011 they reached €200 million, and since 2013 they have averaged €300 million per year (INAIL 2003, 2007, 2014b, 2016, 2017). Art. 10 of Legislative Decree 81 2008 (Repubblica Italiana 2008) has also preserved the responsibility of the enforcing institution, SPSAL, to facilitate intra-institutional cooperation, provide training, counselling activities and published material, and assist and educate workers and employers.

Therefore, the increased funding available to INAIL suggests its central role in promoting OHS in firms. With the latest reforms, the organisation of the Italian deterrence- and compliance-driven OHS enforcement activities have been implemented in accordance with civil and penal code procedures, but also as recommended by the 1989 EEC Council Directive (89/391/EEC) on OHS.[19] In other words, only recently has the Italian OHS enforcement system been reformed to allow a coherent implementation of a due-process legal system while combining deterrence-driven with compliance-driven enforcement activities.

Since the 1990s, the Italian institutions involved in preventing health and safety incidents have been designed to implement a deterrence and

---

[19]To give an example, in January 2013, Milan's SPSAL organised a conference to discuss the risk involved in nursing homes. This led to the participation of 90 professionals including SPSAL directors, safety managers, medics, employees' safety representatives and technicians of 65 nursing homes cooperatives and independent firms. The conference was centred on biomechanical injuries caused by the movement of patients in care and led to the publication of various speeches delivered during the conference on the ASL website (Regione Lombardia and Dipartimento Di Prevenzione Medico 2013a). Milan's SPSAL also published reports on the enforcement activities conducted to curb the issue of agricultural works undertaken in urban areas. The reports illustrate the issues encountered by employers and workers during these working activities, and the major issues discovered during enforcement that can be used by inspectors when planning visits (Regione Lombardia and Dipartimento Di Prevenzione Medico 2013b).

information system aimed at achieving compliance, reducing incidents and enforcing the law consistently. A major critique of the Italian health and safety information database, which INAIL became responsible for maintaining in 2000 (Repubblica Italiana 2008), is that it has only been made available since the mid-2010s, which represents a long delay if compared to the HSE web database and information portal.

2.3   Penalties and Sentences

The number of penalties issued, and the total amount paid by employers due to OHS crimes in any given year are not recorded centrally by the Italian government, and therefore it is only possible to estimate the level of penalties issues and collected from employers. Despite this issue, it is possible to make several speculations. Since 1994, penalties for OHS crimes in Italy have been lower but more frequent than in Britain.

Italian penalties are divided between administrative on-the-spot penalties and sentences following prosecutions. Despite this, according to the legislation passed since 2008, it is possible to observe that in real terms, penalty levels have fallen since 1994. Legislative Decree 81 2008 introduced penalties reaching up to €15,000 and a maximum prison sentence of eight months (Repubblica Italiana 2008; Rinaldi 2012; Rausei 2008). However, from August 2009, the centre-right government managed to pass Legislative Decree 106 2009 (Repubblica Italiana 2009), which imposes a maximum penalty of €6400 and a six-month prison sentence for OHS crimes. These penalty levels are similar to those introduced in 1994, which means that effectively, penalties have fallen in real terms.

Since 2001 the Italian legal system has adjusted to the international corporate manslaughter legislation and allowed judges to penalise and sentence firms rather than just people within them (Rinaldi 2012; Rausei 2008); however, the legislation makes a distinction between hazardous firms and non-hazardous ones.[20] Hazardous firms can be

---

[20]Firms handling dangerous chemical materials, defined as thermoelectric plans, adopting nuclear technology, producing explosives, with more than 200 employees, and operating in the extractive industry with more than 50 employees.

penalised up to €1.5 million. For firms detected without risk assessments, the total penalty can reach a maximum of €750,000 (Repubblica Italiana 2001, Art 25-Septise). In both cases, the total penalty cannot exceed the capital or the total assets of the organisation. When deciding the penalty, the judge must take into consideration different aspects, such as the seriousness of the offence, the level of responsibility of the organisation and the extent to which the risk has been eliminated or minimised following the incident. The penalty can also be reduced by half if the firm has gained little or no advantage from it or if the financial damage caused was small. The penalty can be reduced by between a half and one third of the maximum, if adequate compensation has been provided to victims. If both of these extenuating circumstances apply, the penalty can be reduced by between a half and two-thirds, but cannot be lower than €10,000 (Repubblica Italiana 2001). In other words, penalties in Italy are designed to be reduced significantly if duty holders or firms demonstrate a willingness to reform, and if they compensate victims.

In addition, if the legal breach that caused the incident was caused in order to gain profits by people in managerial positions, as a result of a managerial shortcoming, or was recidivous, the court can impose an additional penalty. Courts sentences can also include revocation of licenses and disqualification of managers and directors from their positions. They can also ban firms from bidding for government contracts, and can be excluded from receiving—or revoking existing—state allowances, funding, contributions or aid. Finally, judges can temporarily or permanently prohibit firms from advertising company products, or ask them to announce their sentence in newspapers or journals as ordered by the judge (Repubblica Italiana 2001, Art. 25-septies, 11, 12, 13 and 9; Fooks et al. 2007). However, these penalties cannot be applied if the firm has provided adequate compensation for victims, has removed the harmful consequences of the offence before the sentence, and has allowed the profit made from breaches to be expropriated. Failure to comply with court orders results in a prison sentence of six months to three years for firms' directors, and firm's expropriations (Repubblica Italiana 2001). Therefore, it might be argued that the Italian court sentences can result in high penalties but shorter prison sentences when compared to the British ones, especially since the publication of the new 2013 sentencing guideline. On-the-spot administrative penalties are higher and might have a higher impact than the British FFI policy introduced in 2012. Also, the penalties and sentences designed for firms, rather than individuals, are low and allow a wide range of avenues to reduce penalties to as little as €10,000.

3   Summary

OHS regulations have been introduced during the various industrialisation processes that occurred across jurisdictions in Europe in the nineteenth and twentieth centuries. In both Britain and Italy, the level of regulation and the enforcement institutions responsible for enforcing the law have continued to grow in size and scope in order to prevent the social problems caused by a capitalist mode of production and, thus, improve levels of social justice and equality. A major reform occurred in Britain in 1974 when the first OHS goal-setting regulation was adopted, which aims to achieve regulatory outcomes rather than prescribe regulatory processes. This new philosophy was adopted across Europe with the ECC Directive 89/391/ECC, and in Italy introduced with the Legislative Decree 626 1994 (Repubblica Italiana 1994a). By 1979 Carson's (1979) was the first to observe that the OHS regulation and enforcement institutions' responses to these crimes were inconsistent when compared to equally harmful crimes. Carson argued that OHS crimes were subject to under-criminalisation and decriminalisation. This was the case despite the increasing scope of the regulation and the size that the enforcement institutions had reached by then. While until the 1990s in both Britain and Italy the growth of enforcement institutions followed similar patterns, since then the expansion and scope of the OHS enforcement institutions in Britain has stopped and, perhaps, reversed.

Since 2004, the HSE enforcement officers have witnessed a constant reduction in their capabilities to enforce the regulations with deterrence-driven policies. British governments have also reduced the funding available to the institution. Arguably, this has been done with the intention to force the OHS enforcement institutions to use compliance-driven policies. These are more financially efficient to implement than the traditional deterrence-driven ones and can function as a relief for businesses, but further decriminalise OHS crimes. The 2008 global financial crisis has given British governments a renewed justification to reform the regulatory regime in order to boost the national economy, but this has further decriminalised OHS crimes and increased social injustice and inequality.

Various political rhetoric has been used in Italy, but this has not been as effective as in Britain. Due to constitutional values which are designed to promote[21] and, hence, guarantee citizens' rights, the decriminalisation of OHS crimes has been much more limited for three main reasons. First, the Italian Constitution guarantees a minimum level of law enforcement, including that of OHS. Since at least the 1990s, 5% of national firms annually has been considered the minimum level of enforcement. Second, it also imposes a strong separation of powers between the state executive, legislative and judiciary. If compared to British political model, the Italian constitutional-driven separation of powers doctrine gives much more authority to Parliament to decide whether to pass legislation, reduce public funding, or change OHS enforcement policies. Third, the traditional design of the Italian criminal justice system, which abides strictly to due-process principles of legal consistency, proportionality and fairness, has also prevented this decriminalisation process. Compared to Britain, the procedures dictating the decisions to prosecute health and safety crimes must abide to the CP and CPP, which is designed to create persistent criminal justice system responses across criminal. The CP and CPP attempt to match the social harm caused by criminalised acts to the responding punishment of the criminal justice system and the procedures by which the investigation and prosecution might start. This is not as strict in Britain, where criminal justice system responses to crimes, albeit increasingly rationalised by the CPS, are not standardised across *all* deviant actions.

The organisation and responsibilities of the enforcement authorities in both jurisdictions differ. The main British enforcement institution is the

---

[21]The Italian Constitution is said to be a promotional constitution (costituzione propositiva) because it encourages the promotion of people's civil rights.

HSE, a nationwide executive non-departmental public body associated with the DWP, for which accountability and responsibility for the achievement of its statutory duty is assigned to the SoSWP. LAs also have OHS enforcement responsibilities, but mainly towards the service sector. LAs are expected to follow HSE enforcement guidelines decided by the HSE and SoSWP through liaisons committees and panels. Both organisations are very financially efficient in enforcing the regulation with deterrence-driven and compliance-driven policies. The independence of the HSE from the government is limited due to the SoSWP's involvement in organisational decisions. This institutional organisation is very cost-effective, but raises concerns in terms of regulatory capture and the conflict of interests that enforcement officers might encounter during their field activities.

In Italy, ITLs, directorates are subordinate to the Ministry of Labour and Social Policies, and have traditionally been responsible to enforce labour laws, including OHS. In 1980, the SPSALs, organisations subordinate to the Ministry of Health and the Regional Health Service, became the main OHS enforcement authority. Both enforcement institutions must coordinate enforcement activities jointly and in accordance to the CP and the CPP. Hence, Italian Ministers have less power to affect OHS policies than the SoSWP in Britain. INAIL is the national compulsory OHS insurance administration for all businesses employing workers. The insurance scheme incentivises the reporting of incidents and, thus, it is also responsible for collecting incidents statistics. Since the early 2000s, INAIL has increasingly acquired a regulatory promotional role by distributing grants to SMEs aiming at improving OHS in the economy. This, effectively, represents compliance-driven law enforcement policy. In years to come, SPSALs traditional regulatory promotional role may increasingly be transferred to INAIL. This will greatly reduce the conflict of interest that SPSALs officers experience while both enforcing the regulation through deterrence-driven policies and while offering support to firms to achieve regulatory compliance through compliance-driven ones. Thus, the difference between the British and Italian OHS enforcement institutions organisation and responsibilities is also affected according to the political relationships between central and local administrations, devolution of political powers and criminal justice system enforcement models and procedural codes.

Enforcement activities also differ significantly between the two jurisdictions. Since the early 2000s, British enforcement policies have increasingly shifted from deterrence- to compliance-driven approaches. These will be analysed in greater depth in the following chapter. The continuous reduction of funding to the HSE has significantly reduced the organisation's capacity to conduct proactive and reactive enforcement activities aiming at penalising and prosecuting non-compliers. The 2012 FFI policy has allowed the HSE to raise revenues and partially repair this issue. Since 1994, SPSALs' enforcement activities have conformed to national enforcement practices as dictated by the CP and CPP. This has stressed SPSALs' responsibility to achieve regulatory compliance in firms through deterrence-driven strategies rather than by providing unpaid-for or non-penalising advice and support. The annual proactive Italian OHS enforcement target has remained unchanged at 5% of firms in the territory since the 1990s, but waves of budget cuts introduced since then have at times impaired enforcement officers' activities. Chapter 5 will analyse this issue in greater depth.

The level and types of penalties issues for non-compliance also differ radically between Britain and Italy. British OHS enforcement officers did not have the power to issue any on-the-spot penalising charges to non-compliers until 2012, when they were given the power to charge duty holders for the technical support provided while encountering material (i.e. significant) breach of the regulations. Italian SPSALs officers have been issuing on-the-spot fixed *compulsory* penalties since 1994. The most interesting differences between the two jurisdictions concern the reasons why judicial on-the-spot administrative penalties cannot be issued in Britain but represent a standardised routine in Italy, and why British enforcement officers can charge firms for providing support, which is not different from a consultancy service, but Italian jurists consider such act a profanation of the fundamental principles of the Italian criminal justice system. The British judiciary perceives any on-the-spot administrative penalty issued by the enforcement institutions as a threat to citizens' liberties and an abuse of power by the government. The HSE FFI policy was introduced to obviate the lack of on-the-spot penalties and persuade duty holders to actively seek compliance *before* officers' inspections, rather than only after. Italian law enforcement agencies cannot charge a duty holder for their services because it would question the SPSALs' institutional independence and create conflicts of interests between the enforcement officers' primary objective, which is to achieve greater compliance among duty holders, and the criminal justice system's main objective of achieving justice. The Italian OHS enforcement institutions' organisation and powers, in other words, are designed to avoid this conflict and achieve social justice consistently, proportionately and, hence, fairly. This is less the case in Britain (and especially in England) because the penalties issued are not strictly regulated by an Italian equivalent of the CP, and enforcement policies are also affected by the SoSWP's decisions.

British and Italian enforcement institutions' decisions and procedures to prosecute also differ significantly, but in both jurisdictions OHS crimes are under-criminalised. In Britain, the decision to prosecute is taken by the enforcement institution as guided mainly by principles of the Code of Crown Prosecution, which are the availability of evidence, a realistic prospect of conviction and the public interest test. In Italy, according to the CPP, cases classified as prosecutable are automatically referred to public prosecutors within the PM, whose decisions are independent from the government and based only on whether there is enough evidence to proceed with the court trial. Also, HSE and LAs can take cases to court directly, without transferring the case to the police or the CPS (except in special situations, such as for manslaughter cases), which means that the institution is also responsible for paying for the financial costs of court trials. The HSE has a keen interest to maximise the prosecution rate of OHS crimes in order to use resources efficiently. In Italy, this cannot happen at enforcement institutions stage, and at the moment it is still difficult to understand the prosecution rate of OHS court cases because statistics are not available. However, according to an Italian enforcement officer, the rate of prosecutions achieving conviction is about 50%, which is much lower than the 90%-plus in Britain. The issues in Italy are caused by the public prosecutors' and judges' lack of technical knowledge in OHS matters. Hence, while in Britain under-criminalisation is caused by the enforcement institutions practices and policies, in Italy the under-criminalisation happens only during court trials and due to the inability to reach guilty setnences. (Table 1).

Table 1 Principal OHS policy landmarks in Britain and Italy

	Great Britain	Italy
1833	The Factory Act 1833 institutes the Factory Inspectorate, which is given responsibility to enforce the first British health and safety law. Penalties can only be issued after magistrate court prosecution. By the 1840s the Factory Inspectors start using compliance-driven policies because of difficulties winning court cases against employers	
1899		First health and safety laws are enacted (RD 230/1/2 1899). The Labour Inspectorate is given responsibility to enforce the health and safety regulations
1933		INAIL is instituted (RD 264 1933)
1947		Italian Constitution is approved. Article 109 institutes the UPG
1955/6		The Italian occupational health and safety regulations are adapted to Italian constitutional principles. DPR 520 1955 and DPR 547 1955, DPR 303 1956. The Labour Inspectorate maintains health and safety enforcement responsibilities
1974	HASW Act 1974 is approved and introduces a tripartite system and goal-setting philosophy based on the concept of ensuring the health and safety of workers as far as is reasonably practicable. The '974 Act also institutes the Health and Safety Commission (HSC) and HSE with the responsibility—together with Las—to enforce regulations	

(continued)

Table 1 (continued)

	Great Britain	Italy
1980		The Italian National Health Service Law 833 1978 institutes the SPSAL, which take over most of the health and safety enforcement responsibility from the Labour Inspectorate. However, SPSAL officers do not have full UPG powers, which means that cannot issue on-the-spot penalties, improvement or prohibition notices and prosecute businesses without the support of DPL. The SPSAL can use compliance-driven policies to enforce regulations, hence by using suggestions and helping businesses
1981		The De-penalisation Law 689 1981 introduces improvement and prohibition notices but only for UPGs, which means that SPSAL officers are not obliged to use it
1989	EEC Council Directive (89/391/EEC) is approved. This introduces a goal-setting philosophy based on the principle of *technological viability*. The Directive guarantees a minimum level of health and safety standards across the European Economic Community. The directive does not prescribe how to enforce regulations, and allows member states to institute enforcement services as well as educational services for businesses. The Directive does not prescribe whether the enforcement and educational practices can be conducted by the same institution. From this point onward the health and safety regulations of the member states will need to be approved at European level, and the only way to release businesses from the burden of health and safety regulations can be done by changing the enforcement policies	

(continued)

Table 1 (continued)

	Great Britain	Italy
1994		DL 626 1994 introduces a goal-setting philosophy based on the concept of ensuring the health and safety of workers as far as it is *technologically viable*. The DL 758 1994 confers SPSAL officers UPG powers. From this point they are obliged to issue on-the-spot administrative penalties for every health and safety breach detected, issue improvement and prohibition notices, and report prosecutable breaches and incidents to the PM. SPSAL officers lose the discretionary powers enjoyed since 1980. Enforcement officers can use compliance-driven policies only when not inspecting firms or investigating incidents
		INAIL start acquiring responsibilities to promote health and safety in firms. SPSAL starts gradually losing funds not dedicated to deterrence-driven enforcement activities
2000	The 2000 *Revitalising health and safety* report suggest introducing enforcement policies that emphasise much more workers' health and safety rights	
2004	The 2004 *Strategy for workplace health and safety in Britain to 2010 and beyond* represents a U-turn in terms of OHS enforcement policies. From this point onward the British OHS enforcement policies were gradually weakened to favour businesses and deprive workers of their rights. A *House of Commons Work and Pensions Select Committee* report urges the HSE and the government to start introducing the policies suggested in the 2000 report, but the request was not followed up. As a consequence, the HSE starts losing funds and front-line enforcement officers	

(continued)

Table 1 (continued)

	Great Britain	Italy
2005	The ECJ takes Great Britain to court arguing that the *reasonable practicability* principle used in the HASW Act 1974 is incompatible with the *technologically viable* principle used in the 1989 EEC Directive. In 2007 the case is dropped and Great Britain is allowed to use the *reasonable practicability* principle	
2005	The *Hampton Report* suggests that resources should be "released from unnecessary inspections [and] redirected towards advice to improve compliance" (Hampton 2005, p. 1). The suggestions are fully endorsed by the government and the HSE	
2007	The Regulators' Compliance Code proposes a "risk-based proportionate and targeted approach to regulatory inspection and enforcement" (Better Regulation Executive 2007, p. 5). This offers a justification to further reduce the amount of proactive HSE proactive (pre-planned) enforcement activities	
2008		DL.ivo 81 2008 is approved. This is a reformed version of the DL.ivo 626 1994. Administrative penalties increase and court sentences more criminalised. INAIL acquire even more power to promote occupational health and safety in the economy through grants and training
2009	The HSE Enforcement Policy Statement is updated. The prospect of prosecution (the only way to penalise businesses for health and safety breaches at the time) become even slimmer	DL.ivo 106 2009 reduces administrative penalties and court sentences issuable back to 1994 levels

(continued)

Table 1 (continued)

	Great Britain	Italy
2010	The Conservative-Liberal Democrat government is elected and announces a reduction of the HSE budget by 35% or £84 million between 2011 and 2015	L 122 2010 transferred to INAIL even more responsibilities to promote occupational health and safety in the economy with grants, conferences and by creating a digital database
2011	The 2011 Department of Work and Pensions report Good Health and Safety, Good for Everyone suggests "a very substantial drop in the number of health and safety inspections carried out in the UK" (Department of Work and Pensions 2011a, p. 3)	
2011	The *Löfstedt* report suggests that the government resolve the health and safety compensation culture developed since 2000. As a consequence, workers' rights to demand compensation for health and safety incidents are reduced, despite the fact that between 2000 and 2011 the amount of health and safety compensation has fallen by 60%. This leads in 2013 to the abolition of the No-Win-No-Fee legal subsidies allowing workers to seek compensation from employers without risking financial losses	

(continued)

Table 1 (continued)

	Great Britain	Italy
2012	The Health and Safety (Fees) Regulation 2012 introduces a charge for firms for material breaches, or when the enforcement officers are required to issue suggestions and notices. This effectively introduces a deterrence-driven enforcement policy designed to deter firms from breaching the health and safety law before officers' inspection. By 2013, however, the amount of charges collected is only £5.5 million, much less than £84 million reduction of the HSE budget announced in 2010	
2014		New Sentencing Guideline 2013 is introduced. Various Prime Minister and Ministerial Decrees rationalise the organisation of the labour inspectorate, including health and safety, and enforcement activities, and organise it in a stricter vertical hierarchical structure. The PSRC is still responsible to take strategic health and safety enforcement decisions. The Central and Interregional Coordination Commissions and Territorial Labour Directorates are responsible to execute these directives on the territory

References

Almond, P. (2013). *Corporate manslaughter and regulatory reform*. Basingstoke: Palgrave Macmillan.

Baldwin, R., & Veljanovski, C. G. (1984). Regulation by cost-benefit analysis. *Public Administration, 62*(Spring), 51–69.

Bartrip, P. W. J., & Fenn, P. (1980a). The conventionalization of factory crime a re-assessment. *International Journal of the Sociology of Law, 8,* 175–186.

Bartrip, P. W. J, & Fenn, P. (1980b). The administration of safety: The enforcement policy of the early factory inspectorate 1844–1864. *Public Administration, 58*(Spring), 87–102.

Bartrip, P. W. J., & Fenn, P. (1983). The evolution of regulatory style in the nineteenth century British factory inspectorate. *Journal of Law and Society, 10*(1983), 201–222.

Better Regulation Executive. (2007). *Regulators' Compliance Code. Statutory Code of Practice for Regulators*. London: Department for Business Enterprise & Regulatory Reform, 2007 [Online]. Available from: http://www.berr.gov.uk/files/file45019.pdf. Last Accessed 25 Oct 2013.

Bonomi, B., & Marinaro, M. (2009). *Dossier sicurezza. Salute e sicurezza sui luoghi di lavoro: Le strategie di prevenzione degli infortuni sul lavoro e di promozione dei livelli di salute e sicurezza sul lavoro*. Carsoli: Ministero del Lavoro, della Salute e delle Politiche Sociali and il Sole 24 S.p.A [Online]. Available from: http://www.lavoro.gov.it/SicurezzaLavoro/Documents/Dossier_sicurezza_web_EXE.pdf. Accessed 25 Aug 2014.

Carson, W. G. (1979). The conventionalisation of early factory crime. *International Journal of the Sociology of Law, 7,* 37–60.

Cataldi, E. (1983). *L'Istituto Nazionale per l'Assicurazione contro gli Infortuni sul Lavoro: testimonianza di un secolo*, INAIL 1983 [Online]. Available from: http://www.inail.it/internet/default/INAILcomunica/ListaPubblicazioni/p/DettaglioPubblicazioni/index.html?wlpnewPage_contentDataFile=UCM_PORTSTG_104030&wlpnewPage__dettaglioDaArchivio=true&_windowLabel=newPage. Accessed 25 Aug 2014.

Coordinamento Tecnico Interregionale. (2011). *Attività delle regioni e delle province autonome per la prevenzione nei luoghi di lavoro, elaborazione PREO as reported by Conferenza delle regioni e delle provincie autonome* [Online]. Available from: http://www.regione.emilia-romagna.it/sicurezza-nei-luoghi-di-lavoro/coordinamento/altre-strutture-e-documenti-di-riferimento/piani-nazionali-e-regionali. Accessed 25 Aug 2014.

Council of Europe. (2010, December 17–18). *European standards as regards the independence of the judicial system: Part II—The prosecution service.* Study N° 494/2008, CDL-AD(2010)040, January 3. Adopted by the European Commission For Democracy Through Law at its 85th plenary session (Venice) [Online]. Available from: http://www.venice.coe.int/webforms/documents/?pdf=CDL-AD(2010)040-e. Accessed 20 Mar 2018.

Davis, C. (2004). *Making companies safe: What works?* Centre for Corporate Accountability [Online]. Available from: http://www.unitetheunion.org/uploaded/documents/Making%20Companies%20Safe%20-%20what%20works%20(CCA-Unite%20paper)11-4856.pdf. Accessed 25 Aug 2014.

Dawson, S., Willman, P., Clinton, A., & Bamford, M. (1988). *Safety at work: The limits of self-regulation.* Cambridge: Cambridge University Press.

Delle Fave, C. (2013). *Manuale di polizia giudiziaria.* Santarcangelo di Romagna: Maggiole Editore.

Department of the Environment, Transport and the Regions (DETR). (2000). *Revitalising health and safety.* Wetherby: Department of the Environment, Transport and the Regions [Online]. http://www.hse.gov.uk/statistics/pdf/prog2009.pdf. Accessed 25 Aug 2014.

Department of Work and Pensions. (2011a). *Good health and safety, good for everyone* [Online]. Available from: https://www.gov.uk/government/publications/good-health-and-safety-good-for-everyone. Accessed 25 Aug 2014.

Department for Work and Pensions. (2011b, November 28). *Reclaiming health and safety for all: Professor Löfstedt's independent review of health and safety legislation and the government response* [Online]. Available from: https://www.gov.uk/government/uploads/system/uploads/attachment_data/file/66790/lofstedt-report.pdf. Accessed 25 Aug 2014.

Dubini, R. (2001). *Articolo 2087 del codice civile. L'obbligo del datore di lavoro di attenersi al principio della massima sicurezza tecnologicamente fattibile.* Sicurezza tecnica, organizzativa e procedural [Online]. Available from: http://www.dbworld.it/file/studi/2087_1329822805.pdf. Accessed 25 Aug 2014.

Dugan, E. (2013, July 30). Compensation culture is a myth: Claims for work-related injuries and diseases fall 60 per cent in a decade. *The Independent* [Online]. Available from: http://www.independent.co.uk/news/uk/home-news/compensation-culture-is-a-myth-claims-for-workrelated-injuries-and-diseases-fall-60-per-cent-in-a-decade-8738679.html. Accessed 25 Aug 2014.

European Commission. (2005). *Labour inspection (health and safety) in the EU (25 member states)—A short guide.* DG Employment, Social Affair and Equal Opportunities [Online]. Available from: http://ec.europa.eu/employment_social/health_safety/slic_en.htm. Accessed 18 Nov 2008.

European Commission. (2007). *Improving quality at work: Community strategy 2007–2012 on health and safety at work*. Communication from the Commission to the European Parliament, the Council, the European Economic and Social Committee and the Committee of the Regions. COM(2007) 62 [Online]. Available from: http://eur-lex.europa.eu/legal-content/EN/TXT/?uri=celex%3A52007DC0062. Accessed 30 Jan 2018.

European Commission. (2012). *Enforcement of fundamental workers' rights*. Directorate-General for Internal Policies of the Union. ISBN: 978-92-823-3831-5. https://doi.org/10.2861/1781 [Online]. Available from: https://publications.europa.eu/en/publication-detail/-/publication/2b47fb86-73eb-4354-b5ef-996c3d461e25. Accessed 30 Jan 2018.

Fioravanti, M. (2011). Le dottrine dello stato e della costituzione. In R. Romanelli (Ed.), *Storia dello Stato Italiano dall'unità ad Oggi*. Donzelli: Italy.

Fooks, G. (2008). *The relationship between the levels of fines imposed upon companies convicted of health and safety offences resulting from deaths, and the turnover and gross profits of these companies*. Centre for Corporate Accountability [Online]. Available from: http://www.corporateaccountability.org.uk/dl/manslaughter/reform/ccasentresearchmar08.doc. Accessed 25 Aug 2014.

Fooks, G., Bergman D., & Rigby, B. (2007). *International comparison of (a) techniques used by state bodies to obtain compliance with health and safety law and accountability for administrative and criminal offences and (b) sentences for criminal offences*. Health and Safety Executive [Online]. Available from: http://www.hse.gov.uk/research/rrhtm/rr607.htm. Accessed 25 Aug 2014.

Great Britain. (1974). *Health and Safety at Work etc. Act 1974 (c. 37)*. London: HMSO.

Great Britain. (2012). *Health and Safety (Fees) Regulation 2012 (c. 255)*. London: HMSO.

Gunningham, N. (2007). Corporate environmental responsibility: Law and the limits of voluntarism. In D. McBarnet, A. Voiculescu, & T. Campbell (Eds.), *The new corporate accountability: Corporate social responsibility and the law*. Cambridge: Cambridge University Press.

Hampton, P. (2005, March). *Reducing administrative burden: Effective inspections and enforcement*. HM Treasury [Online]. Available from: http://www.fera.defra.gov.uk/aboutUs/betterRegulation/documents/hamptonPrinciples.pdf. Accessed 25 Aug 2014.

Hawkins, K. (2002). *Law as last resort: Prosecution decision-making in a regulatory agency*. Oxford: Oxford University Press.

Hazards Magazine. (2005, August). Protection racket. *Hazards Magazine*, Issue 91 [Online]. Available from: http://www.hazards.org/commissionimpossible/protectionracket.htm. Accessed 25 Aug 2014.

Hazards Magazine. (2010, October–December). Get shirty. *Hazards Magazine,* Issue 112 [Online]. Available from: http://www.hazards.org/votetodie/get-shirty.htm. Accessed 25 Aug 2014.

Hazards Magazine. (2015). *Give up. What can you do when a watchdog just sucks?* [Online]. Available from: http://www.hazards.org/votetodie/giveup. htm. Accessed 15 June 2015.

Hill, A. (2006, November 22). Memorial service for Aberfan disaster. *The Observer* [Online]. Available at: http://observer.guardian.co.uk/uk_news/story/0,,1928511,00.html. Accessed 25 Aug 2014.

House of Commons Work and Pensions Select Committee. (2004). *The work of the Health and Safety Commission and executive.* Fourth Report of Session 2003–04, Vol. I, HC 456–1. London: The Stationery Office [Online]. Available from: http://www.publications.parliament.uk/pa/cm200304/cmselect/cmworpen/456/45602.htm. Accessed 25 Aug 2014.

HSC. (2004a, April 6). *Becoming a modern regulator.* Health and Safety Commission Paper HSC/04/53 [Online]. Available from: http://www.hse.gov.uk/aboutus/meetings/hscarchive/2004/060404/c53.pdf. Accessed 25 Aug 2014.

HSC. (2004b). *Highlights from the HSC annual report and the HSC/E accounts 2003/2004* [Online]. Available from: http://www.hse.gov.uk/aboutus/reports/index.htm. Accessed 25 Aug 2014.

HSC. (2005a). *Sensible health and safety at work: The regulatory methods used in Great Britain.* An account of the approach of the Health and Safety Commission [Online]. Available from: http://www.hse.gov.uk/aboutus/strategiesandplans/sensiblehealthandsafety.pdf. Accessed 25 Aug 2014.

HSC. (2005b). *Highlights from the HSC annual report and the HSC/E accounts 2004/2005* [Online]. Available from: http://www.hse.gov.uk/aboutus/reports/index.htm. Accessed 25 Aug 2014.

HSE. (1998). *Health and Safety (Enforcing Authority) Regulations 1998: A–Z guide to allocation* [Online]. Available from: http://www.hse.gov.uk/foi/internalops/og/og-00073-appendix1.htm. Accessed 18 June 2013.

HSE. (2003). *Regulation, enforcement, inspection and what we will do.* Paper presented to the HSE board, October [Online]. Available from: http://www.hse.gov.uk/aboutus/meetings/hsearchive/2003/030903/item7.pdf. Accessed 25 Aug 2014.

HSE. (2004). *A strategy for workplace health and safety in Britain to 2010 and beyond* [Online]. Available from: http://www.hse.gov.uk/aboutus/strategiesandplans/strategy2010.pdf. Accessed 25 Aug 2014.

HSE. (2012a). *HSE annual report and accounts 2011/2012* [Online]. Available from: http://www.hse.gov.uk/aboutus/reports/index.htm. Accessed 25 Aug 2014.

HSE. (2012b). *Fee for intervention: What you need to know* [Online]. Available from: http://www.hse.gov.uk/pubns/hse48.pdf. Accessed 10 Dec 2012.

HSE (2013a) *How HSE enforces health and safety.* [Online]. Available from: http://www.hse.gov.uk/enforce/enforce.htm#enfpen. Accessed 25 Aug 2014.

HSE. (2013b). *Enforcement management model* [Online]. Available from: http://www.hse.gov.uk/enforce/emm.pdf. Accessed 25 Aug 2014.

HSE. (2013c). *National local authority enforcement code health and safety at work England, Scotland & Wales* [Online]. Available from: http://www.hse. gov.uk/lau/national-la-code.pdf. Accessed 23 Feb 2018.

HSE. (2014). *HSE and Local Authority (LA) regulators working together* [Online]. Available from: http://www.hse.gov.uk/laU/index.htm. Accessed 25 Aug 2014.

HSE. (2015a). *HSE principles for Cost Benefit Analysis (CBA) in support of ALARP decisions* [Online]. Available from: http://www.hse.gov.uk/risk/theory/alarpcba.htm. Accessed 15 June 2015.

HSE. (2015b). *Cost Benefit Analysis (CBA) checklist* [Online]. Available from: http://www.hsc.gov.uk/risk/theory/alarpcheck.htm. Accessed 15 June 2015.

HSE. (2015c). HSE *enforcement policy statement* [Online]. Available from: http://www.hse.gov.uk/pubns/hse41.pdf. Accessed 31 Jan 2018.

HSE. (2016). *HSE annual report 2015/16* [Online]. Available from: https://www.gov.uk/government/uploads/system/uploads/attachment_data/file/534093/hse-annual-report-and-accounts-2015-2016.pdf. Accessed 3 Mar 2017.

HSE. (2017). *HSE annual report and accounts 2016/17* [Online]. Available from: http://www.hse.gov.uk/aboutus/reports/ara-2016-17.pdf. Accessed 3 Mar 2017.

HSE. (2018a). *Local authority enforcement* [Online]. Available from: http://www.hse.gov.uk/lau/enforcement.htm. Accessed 23 Feb 2018.

HSE. (2018b). *Advice and guidance* [Online]. Available from: http://www.hse.gov.uk/enforce/advice-information-guidance.htm. Accessed 23 Feb 2018.

INAIL. (2003). *Bilancio Consuntivo 2002* [Online]. Available from: https://www.inail.it/cs/internet/docs/1_bil-cons-2002-pdf.pdf. Accessed 6 Apr 2018.

INAIL. (2007). *Bilancio Consuntivo 2006* [Online]. Available from: https://www.inail.it/cs/internet/docs/bilancio-cons-2006-pdf.pdf. Accessed 6 Apr 2018.

INAIL. (2014a). *La storia* [Online]. Available from: http://www.inail.it/internet/default/Chisiamo/Lastoria/index.html. Accessed 25 Aug 2014.

INAIL. (2014b). *Bilancio Consuntivo 2013* [Online]. Available from: https://www.inail.it/cs/internet/docs/all-bilancio-consuntivo-2013.pdf. Accessed 6 Apr 2018.

INAIL. (2016). *Bilancio Consuntivo 2015* [Online]. Available from: https://www.inail.it/cs/internet/docs/ammt-bilancio-consuntivo-2015.pdf. Accessed 6 Apr 2018.

INAIL. (2017). *Bilancio Previsione 2017* [Online]. Available from: https://www.inail.it/cs/internet/docs/ammt-bilancio-previsione-2017.pdf. Accessed 6 Apr 2018.

Lord Young of Graffham. (2010, October 15). *Common sense, common safety: A report by Lord Young of Graffham: A report by Lord Young of Graffham following a Whitehall-wide review of the operation of health and safety laws.* The Cabinet Office Policy Paper [Online]. Available from: https://www.gov.uk/government/uploads/system/uploads/attachment_data/file/60905/402906_CommonSense_acc.pdf. Accessed 25 Aug 2014.

McBarnet, D., Voiculescu, A., & Campbell, T. (Eds.). (2007). *The new corporate accountability: Corporate social responsibility and the law.* Cambridge: Cambridge University Press.

Ministro Del Lavoro, Della Salute Delle Politiche Sociali. (2008). Decreto 18 settembre 2008 [Online]. Available from: http://www.gazzettaufficiale.it/atto/serie_generale/caricaDettaglioAtto/originario?atto.dataPubblicazioneGazzetta=2008-10-18&atto.codiceRedazionale=08A07492&elenco-30giorni=false. Accessed 30 Jan 2018.

Ministro del Lavoro e delle Politiche Sociali. (2004). *Circolare Ministro del Lavoro e delle Politiche Sociali n.24 del 24 giugno 2004* [Online]. http://www.inps.it/circolariZip/Circolare%20numero%20132%20del%2020-9-2004_Allegato%20n%201.pdf. Accessed 25 Aug 2014.

Mousourakis, G. (2015). *Roman law and the origins of the civil law tradition.* Basel: Springer International Publishing. ISBN 978-3-319-12267-0; e-ISBN 978-3-319-12268-7; https://doi.org/10.1007/978-3-319-12268-7.

Ogus, A. (1994). *Regulation: Legal form and economic theory.* Oxford: Clarendon Press.

Pais, P. R. (2008). *Nuova normativa di tutela e salute sui luoghi di lavoro.* Roma: Epc.

Pearce, D. W. (1983). *Cost-benefit analysis.* London: Macmillan.

Pelliccia, L. (2008). *Il nuovo testo unico di sicurezza sul lavoro.* Rimini: Maggioli Editore.

Porreca, G. (2008). *Istituita con Il D. Lgs. n.124/2004 La Diffida E La Prescrizione Obbligatoria Per Gli Ispettori Del Lavoro. Ma E' Stata Fatta Chiarezza Sull'attivita' Di P. G. In Materia Di Sicurezza Sul Lavoro?* [Online]. Available from: http://www.porreca.it/Presentazione%20decreto%20funzioni%20ispettive.htm. Accessed 2 Oct 2008.

Posner, E. (2003, December). Transfer regulations and cost-effectiveness analysis. *Duke Law Journal, 53*(3), 1067–1079.

Rausei, P. (2006). *Codice delle ispezioni Volume 1 e Volume 2.* Italy: Kluwer Italia.

Rausei, P. (2008). *Vigilanza, ispezioni e sanzioni. La nuova disciplina.* Italy: Wolters Kluwer.

Regione Friuli-Venezia Giulia, Azienda per i servizi Sanitari n.1 Triestina, Struttura Complessa Prevenzione E Sicurezza Negli Ambienti Di Lavoro (SCPSAL). (2013a). *Guida Utile* [Online]. Available from: http://www.ass1.sanita.fvg.it/servlet/page?_pageid=71&_dad=pass1&_schema=PASS1&act=2&id=3684. Accessed 25 Aug 2014.

Regione Friuli-Venezia Giulia, Azienda per i servizi Sanitari n.1 Triestina, Struttura Complessa Prevenzione E Sicurezza Negli Ambienti Di Lavoro (SCPSAL). (2013b). *Guida Utile: Vigilanza Negli Ambienti di Lavoro* [Online]. Available from: http://www.ass1.sanita.fvg.it/servlet/page?_pageid=71&_dad=pass1&_schema=PASS1&act=2&id=786. Accessed 25 Aug 2014.

Regione Friuli-Venezia Giulia, Azienda per i servizi Sanitari n.1 Triestina, Struttura Complessa Prevenzione E Sicurezza Negli Ambienti Di Lavoro (SCPSAL). (2013c). *Guida Utile. Direzione Amministrativa* [Online]. Available from: http://www.ass1.sanita.fvg.it/servlet/page?_pageid=71&_dad=pass1&_schema=PASS1&act=2&id=884. Accessed 25 Aug 2014.

Regione Friuli-Venezia Giulia, Azienda per i servizi Sanitari n.1 Triestina, Struttura Complessa Prevenzione E Sicurezza Negli Ambienti Di Lavoro (SCPSAL). (2013d). *Guida Utile. Direzione Strategica (Staff)* [Online]. Available from: http://www.ass1.sanita.fvg.it/servlet/page?_pageid=71&_dad=pass1&_schema=PASS1&act=2&id=2888. Accessed 25 Aug 2014.

Regione Lombardia, Dipartimento Di Prevenzione Medico, Prevenzione E Sicurezza Negli Ambienti Di Lavoro. (2013a). *Progetto Sanita' Rsa 2011/2012 – Presentazione* [Online]. Available from: http://www.asl.milano.it/ITA/Default.aspx?SEZ=2&PAG=74&NOT=5497. Accessed 25 Aug 2014.

Regione Lombardia, Dipartimento Di Prevenzione Medico, Prevenzione E Sicurezza Negli Ambienti Di Lavoro. (2013b). *Agricoltura E Manutenzione Verde* [Online]. Available from: http://www.asl.milano.it/ITA/Default.aspx?SEZ=2&PAG=74&NOT=5523. Accessed 25 Aug 2014.

Regione Marche, Giunta Regionale Servizio Sanità, Unità O. O. Sanità Pubblica e Prevenzione. (1995). *Ripercussioni Del Decreto Legislativo 758/94 Sulla Operatività Dei Servizi Di Prevenzione Nei Luoghi Di Lavoro* [Online]. Available from: cd494.mannelli.info/files/linee_guida_758.doc. Accessed 25 Aug 2014.

Regione Sicilia, Azienda Sanitaria Provinciale Trapani, Dipartimento di Prevenzione della Salute, U.O. Prevenzione igienico-sanitaria ed epidemiologia occupazionale. (2013). *Assistenza, informazione e formazione* [Online]. Available from: http://www.asptrapani.it/servizi/Menu/dinamica.aspx?idArea=18681&idCat=22333&ID=22334. Accessed 25 Aug 2014.

Regione Veneto, Azienda ULSS 13 Mirano. (2013). *Assistenza, formazione-informazione, promozione della salute* [Online]. Available from: http://www.ulss13mirano.ven.it/nqcontent.cfm?a_id=7668. Accessed 25 Aug 2014.

Repubblica Italiana. (1930). *Reggio Decreto* del 19 ottobre, 1930 n.1398 (e seguenti modifiche). Codice Penale 1930. Gazzetta Ufficiale n.251 del 26 ottobre 1930.

Repubblica Italiana. (1942). *Regio Decreto* del 16 Marzo 1942, n.262 (e seguenti modifiche). *Codice Civile 1942*. Gazzetta Ufficiale n.79, del 4 aprile 1942.

Repubblica Italiana. (1955). *Decreto del Presidente della Repubblica del 27 aprile 1955, n.520*. Riorganizzazione centrale e periferica del Ministero del lavoro e della previdenza sociale. *Gazzetta Ufficiale* n.149 del 1 luglio 1955.

Repubblica Italiana. (1978). *Legge n.833, 23 dicembre 1978,* Istituzione del servizio sanitario nazionale. *Gazzetta Ufficiale* n.360 del 28 dicembre 1978.

Repubblica Italiana. (1988). *Codice di Procedura Penale* (aggiornato al Decreto del Presidente della Repubblica n.447 del 22 settembre 1988), Legge 18 giugno 1955, n.517. *Gazzetta Ufficiale* n.123, 15 maggio 1955.

Repubblica Italiana. (1994a). *Decreto Legislativo n.626, 19 settembre 1994.* Attuazione delle direttive 89/391/CEE, 89/654/CEE, 89/655/CEE, 89/656/CEE, 90/269/CEE, 90/270/CEE, 90/394/CEE, 90/679/CEE, 93/88/CEE, 95/63/CE, 97/42/CE, 98/24/CE, 99/38/CE, 99/92/CE, 2001/45/CE, 2003/10/CE, 2003/18/CE e 2004/40/CE riguardanti il miglioramento della sicurezza e della salute dei lavoratori durante il lavoro. *Gazzetta Ufficiale* n.265 del 12 novembre 1994. Supplemento Ordinario n.141.

Repubblica Italiana. (1994b). *Decreto Legislativo n.758, 19 dicembre 1994.* Modificazioni alla disciplina sanzionatoria in materia di lavoro. *Gazzetta Ufficiale* n.21 del 26 gennaio 1995, Supplemento Ordinario n.9.

Repubblica Italiana (1997) *Decreto Legislativo n.281, 28 Agosto 1997.* Definizione ed ampliamento delle attribuzioni della Conferenza per i rapporti tra lo Stato, le regioni e le province autonome di Trento e Bolzano ed unificazione, per le materie ed i compiti di interesse comune delle regioni, delle province e dei comuni, con la Conferenza Stato - città ed autonomie locali. *Gazzetta Ufficiale* n.202 del 30 agosto 1997.

Repubblica Italiana. (2000). *Decreto Legislativo n.38, 23 febbraio 2000.* Disposizioni in materia di premi dell'Istituto Nazionale per l'Assicurazione contro gli Infortuni sul Lavoro e le malattie professionali (INAIL). *Gazzetta Ufficiale* n.50 del 1 marzo 2000.

Repubblica Italiana. (2001). *Decreto Legislativo n.231, 8 giugno 2001.* Disciplina della responsabilità amministrativa delle persone giuridiche, delle società e delle associazioni anche prive di personalità giuridica, a norma dell'articolo 11 della legge 29 settembre 2000, n.300. *Gazzetta Ufficiale* n.140 del 19 giugno 2001.

Repubblica Italiana. (2007). *Decreto Del Presidente Del Consiglio Dei Ministri 17 dicembre 2007.* Esecuzione dell'accordo del 1° agosto 2007, recante: "Patto per la tutela della salute e la prevenzione nei luoghi di lavoro. *Gazzetta Ufficiale* n.3 del 4 gennaio 2007.

Repubblica Italiana. (2008). *Decreto Legislativo n.81, 9 aprile 2008* (testo aggiornato al 15 ottobre 2010). Attuazione dell'articolo 1 della legge 3 agosto 2007, n.123, in materia di tutela della salute e della sicurezza nei luoghi di lavoro. *Gazzetta Ufficiale* n.101 del 30 Aprile 2008.

Repubblica Italiana. (2009). *Decreto Legislativo n.106, 3 agosto 2009.* Disposizioni integrative e correttive del decreto legislativo 9 aprile 2008, n.81, in materia di tutela della salute e della sicurezza nei luoghi di lavoro. *Gazzetta Ufficiale* n.180 del 5 agosto 2009, Supplemento Ordinario n.142.

Repubblica Italiana (2010). *Legge n.122, 30 luglio 2010.* Conversione in legge, con modificazioni, del decreto-legge 31 maggio 2010, n.78, recante misure urgenti in materia di stabilizzazione finanziaria e di competitività economica. *Gazzetta Ufficiale* n.176 del 30 luglio 2010, Supplemento Ordinario n.174.

Repubblica Italiana. (2012). *Codice di Procedura Civile* (aggiornato al Decreto Legge 22 giugno 2012). Regio Decreto 28 ottobre 1940, n.1443. *Gazzetta Ufficiale* n.253, 28 ottobre 1940.

Repubblica Italiana. (2015). *Decreto Legislativo 14 settembre 2015, n.149.* Disposizioni per la razionalizzazione e la semplificazione dell'attività ispettiva in materia di lavoro e legislazione sociale, in attuazione della legge 10 dicembre 2014, n.183. Pubblicato nella Gazz. Uff. 23 settembre 2015, n.221, S.O. [Online]. Available from: http://www.gazzettaufficiale.it/eli/id/2015/09/23/15G00161/sg. Accessed 30 Jan 2018.

Rideout, R. W. (1989). *Principles of labour law*. London: Sweet & Maxwell.

Rinaldi, M. (2012). *Il procedimento ispettivo*. Italia: Giuffrè Ediotore.

Rubini, G. (2011). *Lavoro in corso. La "disposizione"*. *Diarioprevenzione Magazine*, 20 dicembre 2011. [Online]. Available from: http://www.diario-prevenzione.net/diarioprevenzione/html/modules.php?name=News&file=print&sid=174. Accessed 25 Aug 2014.

Slapper, G. (2000). *Blood in the bank: Social and legal aspects of death at work*. Aldershot: Ashgate.

Taylor, A. J. (1972). *Laissez-faire and state intervention in nineteenth-century Britain*. London: Palgrave Macmillan.

The Crown Prosecution Service. (2013). *Prosecution policy and guidance, the code for crown prosecutors* [Online]. Available from: http://www.cps.gov.uk/publications/code_for_crown_prosecutors/. Accessed 25 Aug 2014.

The Law Commission. (1998). *Consents to prosecution* [Online]. Available from: http://lawcommission.justice.gov.uk/docs/lc255_Consents_to_Prosecution.pdf. Accessed 25 Aug 2014.

Tiraboschi, M., & Fantini, L. (2009). *Il testo unico della salute e sicurezza sul lavoro dopo il correttivo (D.Lgs. n.106/2009)*. Italia: Giuffrè Editore.

Tombs, S. (2016). Making better regulation, making regulation better? *Policy Studies, 37*(4), 332–349.

Tombs, S. (2017). The UK's corporate killing law: Un/fit for purpose? *Criminology and Criminal Justice*, Published online on 23 August 2017 Available from: http://journals.sagepub.com/doi/full/10.1177/1748895817725559. Accessed 25 Mar 2018.

Tombs, S., & Whyte, D. (2007). *Safety crimes*. Cullompton: Willan Publishing.

Tombs, S., & Whyte, D. (2010). *Regulatory surrender: Death, injury and the non-enforcement of law*. Liverpool: Institute of Employment Rights.

UNITE. (2008). *Lack of investigation 2001–2007. Incident reported to the health and safety executive* [Online]. Available from: http://www.uniteth-eunion.org/uploaded/documents/Lack%20of%20Investigation%20of%20Incidents%20reported%20to%20the%20HSE%20(Unite%20report)11-4817.pdf. Accessed 8 May 2018.

Venturi, D. (2014). Prescrizione obbligatoria—Articolo 15. In P. Rausei & M. Tiraboschi (Eds.), *L'ispezione del lavoro, dieci anni dopo la riforma. Id.lgs. n.124/2004 fra passato e future* (ADAPT Professional Series no. 3, pp. 332–343). Modena: ADAPT University Press. ISBN 978-88-98652-28-0.

Woolf, A. (1973). The Robens report—The wrong approach. *Industrial Law Journal, 2,* 88–95.

3

Incidents and Enforcement Trends

Another difference between Britain and Italy is how incidents trends have changed, how these incidents are recorded, and the enforcement activities conducted in Britain and Italy. This chapter attempts to compare British and Italian incidents trends by comparing official figures' incident ratios per workers and major economic activity. No research has attempted to compare occupational health and safety (OHS) incidents trends between two countries due to the complexities of the methodologies used to collect these figures, and the challenges this complexity causes when attempting to create incident ratios per worker and economic sectors. The comparative analysis of incidents and enforcement activities has been conducted with the official data available. To reduce errors, incidents statistics have been clustered in five–seven-year averages for Britain, and three–five-year averages for Italy. Despite the methodological challenges encountered, it is important to attempt this longitudinal comparative analysis because the way incidents are measured in each jurisdiction will also determine the level of social and political alarm this issue cause.

This chapter demonstrates that the most important difference when designing the OHS recording regulation (or methodology) is a political

© The Author(s) 2019
D. Canciani, *The Politics and Practice of Occupational Health and Safety Law Enforcement*, Critical Criminological Perspectives,
https://doi.org/10.1007/978-3-319-98509-1_3

willingness to create an accurate recording system and the stakeholders' incentive to reports incidents. The British recording system creates a little incentive, while the Italian one seems to incentivise incidents reporting much more, which suggests that in Britain there is less commitment to creating a reporting system capable of capturing the real extent and social cost of OHS crimes. A lack of alarm on the issue of OHS harm in Britain was one of the major justifications in the political rhetoric used to implement more compliance-driven enforcement policies and reduce the financial resources given to the HSE, all of which have further decriminalised OHS crimes. This chapter, therefore, aims to analyse and compare the British and Italian incident trends and enforcement activities conducted. Firstly, it analyses the British incidents and enforcement trends, secondly, it analyses the Italian incidents and enforcement trends, and lastly, it compares the two jurisdictions trends by accounting for Tombs and Whyte (2008) levels of incidents mis- and under-reporting.

1   Britain

The Health and Safety Executive (HSE) publishes the injury and enforcement official statistics on its website. The Reporting of Injuries, Diseases and Dangerous Occurrences Regulations (RIDDOR) makes it compulsory to report any injury resulting from "an accident arising out of or in connection with work" that has caused an absence from work for more than seven days, with some exceptions (Great Britain 2013, Art. 2; HSE 2013d). However, it is important to note that the data sets available have significant weaknesses (Tombs and Whyte 2008). The under- and misreporting of incidents seems to be an issue, particularly in Britain. According to Tombs and Whyte (2008) "In order to reconstruct the data on injuries usefully, then, we would need to apply a multiplier of much greater than five to six times HSE headline figure" (Tombs and Whyte 2008, p. 3). Moreover, the reporting ratio changes dramatically between employees and the self-employed who are "twice as likely to be killed [during working activities] but four to five times less likely to sustain a major injury when compared to other workers" (Tombs and Whyte 2008, p. 3).

A major weakness, when compared to Italy, is that fatal and non-fatal road-related working incidents are excluded. RIDDOR require to record some incidents involving vehicles within the workplace premises (see HSE 2013c), but these exclude those incidents occurring to workers who drive for work, such as delivery curriers. The fatal injuries recorded exclude work-related road incidents, which average 1300 deaths and 13,000 injuries per year (HSE 2003a). The HSE argues that the incidents occurring at the wheel of a vehicle during working hours are investigated by the police because they are not caused by OHS breaches (see HSE 2013b for more details). However, according to Royal Society for the Prevention of Accidents (RoSPA) these incidents should be reported by employers and appear in the official HSE figures, which is the established practice in Italy, because these might arise out of or in connection with work activities and strategic management decisions (Johnson 2010; Bibbings 2014).

This issue is also worsened by the methodological inconsistencies used to record occupational health trends and incidents. In Britain, these are recorded from databases, such as SWORD in 1989, EPIDERM in 1993 and OPRA in 1996, THOR from 2002,[1] and IIDB from 2003; death certificates related to occupational health diseases, such as asbestosis; and from the NHS and other institutions, such as hospitals, general practitioners and occupational therapists. British employees working in hazardous firms are required to undertake regular medical visits (HSE 2018). However, general practitioners' and hospital doctors' reporting occurs on a voluntary basis, which decreases the reliability and consistency of the overall figures (Cherry et al. 2000; Centre for Occupational and Environmental Health 2013). In Italy, hospital doctors and GPs are obliged to question patients if any health condition treated might have been caused by work activities, and report any injury or occupational illness to the National Institute for Insurance against Working Accidents (INAIL) (Repubblica Italiana 2015). This

---

[1]Surveillance of Work-Related and Occupational Respiratory Disease (SWORD), Occupational skin surveillance (EPIDERM), Occupational Physicians Reporting Activity (OPRA), The Health and Occupation Research network (THOR).

legal obligation and the Italian National Health Service's commitment to record and evaluate occupational health-related sicknesses seems better than in Britain, but the Italian level of under-reporting is also difficult to assess.

The final weakness of the HSE official statistics concerns the size of the workforce adopted to create the ratio of incidents per 100,000 workers. The HSE argues that until November 2011 the incidents ratios were created by using the Workforce Jobs employment survey for employees and the Annual Population Survey for self-employed data, which raised the issue of potential double counting and, hence, might decrease the number of incidents per worker (HSE 2013a). However, despite this weakness, the HSE website offers no data for the number of employed and self-employed workers used to create the ratio of OHS incidents published on its website. Hence, it is impossible to double check the reliability of the source used by the HSE and the figures used to create the ratio of incidents per workers. Despite this, an analysis of the trends in incidents can still be conducted by using RIDDOR total incidents recorded on the HSE website and workers number data from the Labour Force Survey (LFS).

In terms of OHS incident figures, the data used in this analysis was collected under RIDDOR 1985 and 1995. The difference between these two Acts' recording methodologies is that from 1996, violent acts at work and suicides on railways and other transport system became reportable. The list of major injuries was also expanded to include a broader range of skeletal fractures and amputations and some dislocations. As a consequence of the adoption of RIDDOR 1995, the HSE argues that the injury statistics cannot be compared to those under the 1985 regulation (HSE 2012b). It is also important to note that RIDDOR 2013 came into force in October 2013, but it has not altered the list of reportable incidents (HSE 2013d).

The data set used for the number of workers employed to create the ratio of incidents per 100,000 workers per major economic activity was taken from the LFS. Two main issues were encountered. Firstly, LFS data are collected in yearly quarters, and the number of workers employed in the economy was measured at a specific time on each year included in the analysis. Secondly, the LFS figures include the workers employed in Northern Ireland, but the HSE does not operate there and does not record incidents in that region, which means that those figures were subtracted from the total number of employed workers used to create the ratio.[2]

The second issue encountered was that the Standard Industrial Classification adopted to record incidents changed in 1980, 1992 and 2003. To make the data comparable from 1986, the industrial classifications were clustered under similar headings to form five main classifications: agricultural, hunting, forestry and fishing; extractive and utility supply industries; manufacturing industry; construction industry; and the services industry (see Table 1). RIDDOR has gradually reclassified some of the minor incidents into major ones, which means that non-fatal major injuries might show as increasing throughout the analysis.

1.1   British Incident Trends

1.1.1 British Fatal Incidents

After creating a ratio of work injuries per 100,000 workers, it is possible to make three observations on the OHS incident trends in Britain (see Table 2).[3] The overall number of fatal injuries decreased by 1.1 per 100,000 workers from 1.9 in the 1986–1991 period, to 0.8 deaths per 100,000

---

[2]The LFS data collected to create the annual ratio has been taken from the spring figures every year between 1986 and 1996 and for the period January–March between 1997 and 2009. In order to make the data comparable, the Office of National Statistics produces Person Weight variables to allow the data to be generalisable to the wider national population. The weights used are WEIGHT1 from 1992 to 2000, PWT03 from 2000 to 2006, PWT07 from 2007 to 2009. Thereafter, a cross tabulation between Region of Usual Residence (Row = URESRG from 1986 to 1992 and URSMC from 1992 to 2008) was used to exclude Northern Ireland figures. The dataset was divided also by Industry Division in main Job (Column: INDDIV), to create ration between Basic Economic Activities (layer: INECAC).

[3]In order to make meaningful and accurate analysis of occupational health and safety incident trends since the end of the 1980s, the analysis conducted for fatal injuries in Britain has been done through the use of five- and six-year averages. Although this system might not offer a very accurate representation of any improvements occurring in recent years, it represents a useful tool for analysing longer-term trends, and helps reduce the errors and misinterpretations caused by sharp changes of incidents between years. The periods used for fatal injuries only are 1986/87–1990/91 (1987–1991), 1991/92–1997/98 (1992–1998), 1998/99–2002/03 (1999–2003), and 2003/04–2007/08 (2004–2008).

Table 1 Historical industrial classifications used for the comparative analysis of British and Italian OHS incidents

HSE and ILO Historical industrial classifications	Agriculture, hunting, forestry and fishing	Extractive and utility supply industries	Manufacturing industries	Construction	Service industry: retail, administrative and educational sectors
SIC80	1	1,2	4,5	6	7,8,9,10
SIC92	A & B	C & E	D	F	G,H,I,J,K,L,M,N,O,P, Q
SIC03	A	B,D & E	C	F	G,H,I,J,K,L,M,N,O,P,Q,R,S,T
ISIC-2	1	2,4	3	5	6,7,8,9
ISIC-3	A & B	C & E	D	F	G,H,I,J,K,L,M,N,O,P,Q

SIC 81, SIC 92 and SIC 08 were adopted by the HSE (this is the reason they are underlined) to classify the incidents by industrial classification. However, since 2009 the classification used has changed to ISIC-2 and then to ISIC-3. This table explains how the incidents' data have been grouped together to create longitudinally comparable datasets

Table 2 British fatal occupational safety injuries per 100,000 workers

Year of publication	Agriculture, hunting, forestry and fishing (2)	Extractive and utility supply (3)	Manufacturing	Construction	Service	All industries	All industries including 1300 estimated work-related road incidents (6)
1987 (4) (1)	8.65	2.30	2.31	7.10	0.57	1.72	7.21
1988	9.64	2.46	2.10	7.97	0.65	1.81	7.11
1989	8.66	15.07	2.04	6.89	0.72	2.40	7.52
1990	10.13	2.25	2.35	7 67	0.73	1.85	6.93
1991	9.59	2.00	2.11	6 76	0.79	1.73	6.93
1992 (1)	10.07	2.93	1.60	5.98	0.67	1.51	6.84
1993	8.89	2.51	1.28	6.09	0.72	1.41	6.83
1994	8.50	5.19	1.33	5.30	0.52	1.23	6.62
1995	10.41	1.51	1.17	4.89	0.48	1.11	6.42
1996	9.34	6.13	0.90	4.68	0.44	1.04	6.28
1997 (5)	13.53	3.24	1.23	5.24	0.41	1.14	6.29
1998	10.26	6.58	1.29	4.50	0.41	1.07	6.15
1999	12.64	3.95	1.47	3.70	0.33	0.97	5.98
2000	10.19	2.51	0.91	4.44	0.28	0.84	5.78
2001	13.72	2.67	1.18	5.46	0.42	1.09	5.95
2002	11.63	4.67	1.13	4.16	0.35	0.94	5.80
2003	11.46	1.05	1.06	3.59	0.37	0.84	5.65
2004 (1)	13.59	3.75	0.79	3.40	0.39	0.86	5.61
2005	11.95	0.72	1.17	3.15	0.32	0.81	5.51
2006	9.69	1.87	1.25	2.72	0.34	0.78	5.44
2007	10.54	3.18	1.01	3.47	0.40	0.88	5.54
2008	12.60	2.71	0.94	3.13	0.33	0.82	5.40

Source HSE. The year shown correspond to the year these figures were published (i.e. 2007 = FY2006/07) (1) Standard industrial classification of economic activities used SIC80 for 1987–1991, SIC92 for 1992–2003, SIC03 for 2004–2008. (2) Excludes sea fishing. (3) Includes the number of injuries in the offshore oil and gas industry collected under the offshore installations safety legislation until 1997. These started to be included into the main statistics from 1997. (4) 1987–1996 reported under the RIDDOR 1985. (5) 1997–2008 reported under RIDDOR 1995. (6) The estimated figures are provided by RoSPA for the year 2012/13 (B bbings 2014). These figures should be interpreted with caution because work-related road incidents might have increased or decreased since the 1980s. However, these are the closest estimates to use for comparing British and Italian figures workers in the 2003–2008 period. This means that from the end of the 1980s to the mid-2000s incidents rates fell by 50.8%. However, from the end of the 1990s the improvement in the trend in fatal incidents has slowed down. While between the periods 1986–1991 and 1992–1998 incidents fell by 36.1%, in the periods 1999–2003 and 2004–2008 the rate of improvement fell by 14.7 and 5.6% respectively (see also HSE 2012a).

The improvement was not equal across the economic sectors. The major and steadier improvement in fatal injury rates occurred in the **construction** industry, which from the periods 1987–1991 to 2004–2008 witnessed a fall from 7.3 to 3.2 fatalities per 100,000 workers, or by 56.4%. In the **manufacturing** industry, fatal incidents also fell significantly. The periods 1987–1991 to 2004–2008 witnessed a fall from 2.2 to 1.0 fatalities per 100,000 workers, or by 52.7%. However, the improvement slowed down significantly in the period 1997–2008 when the trend improved by only 10.3%. The number of fatal injuries in the **service** sector also decreased by 42%, that is from 0.7 fatalities per 100,000 workers in the period 1986–1991 to 0.4 in the period 2003–2008.

While fatal incidents fell by 34.7% up until the end of the 1990s, in the 1999–2008 period fatalities increased by 0.9%. The **extractive** industry recorded a sharp fall in fatal incidents up to 1993, but remained stable until 2008. These figures include the 1988 Piper Alpha explosion, which caused the death of 167 workers on a single event, and doubled the 1987–1991 five-year average from 2.3 to 4.8 fatal injuries per 100,000 workers. Hence, by temporarily excluding the Piper Alpha deaths[4] from the figures, the average number of fatal injuries increased from the end of the 1980s. From 1987–1991 to 2004–2008 the number of fatal injuries increased by 8.6% from 2.3 to 2.4 per 100,000 workers. However, in the 1992–1998 period, fatal incidents in the extractive industry increased by 78.1%, before falling to 46.4% in 1999–2003, and by decreasing another 23.1% in 2004–2008. The **agricultural and fishing** industry recorded the highest increase in fatal incidents in the twenty-two year period of 1987–2008. From a five-year average of 9.3 fatalities per 100,000 in the period 1987–1991, this reached an average level of 11.5 fatal injuries per 100,000 workers for 2004–2008. Incidents increased 8.7% in the periods

---

[4]By temporarily excluding the deaths caused by this tragic event from the trends, I have no intention to offend the victims (and people close to them) and would like to apologise in advance if they are.

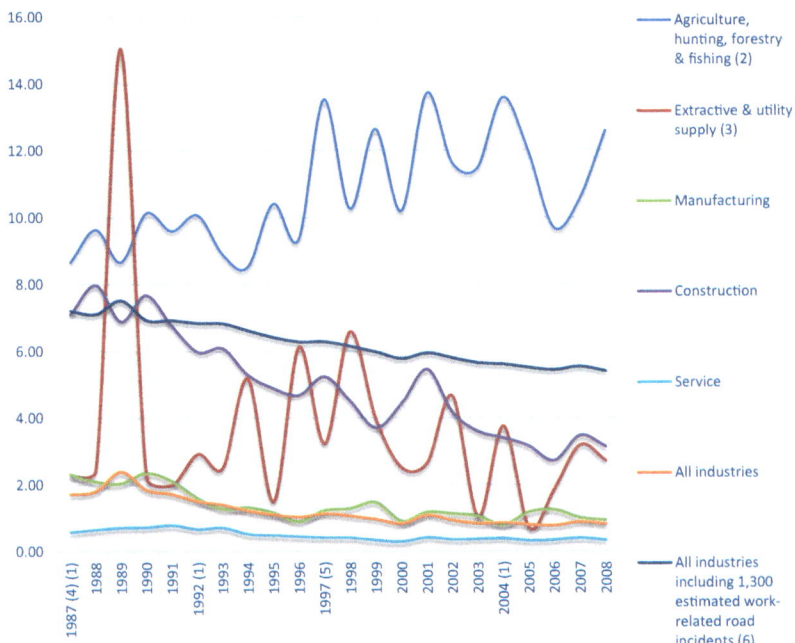

Chart 1 British fatal occupational safety injuries per 100,000 workers (see Table 2 for explanatory footnotes)

1987–1991 and 1992–1998 and by 19.1% in the period 1999–2003, but fell by 2.7% for 2004–2008 (Chart 1).

1.1.2 British Non-fatal Incidents

The situation for non-fatal major injuries in Britain is also analysed (see Table 3). The analysis in this section compares non-fatal major incident trends between 1987–1991 and 1992–1996, and 1997–2003 and 2004–2007. The number of non-fatal major injuries dropped in both the 1987–1996 and 1997–2007 periods. However, while in the 1987–1991 and 1992–1996 periods, incidents decreased by 12.9% from 85.6 to 74.6 per 100,000 workers, in the 1997–2003 and 2004–2007 periods major incidents fell by only 1.3% from 112.4 to 11 per 100,000 workers.

Industrial specific data is also analysed. The sector that on average had the most significant constant fall in major non-fatal incident trends

Table 3 British non-fatal major occupational safety injuries per 100,000 workers

Year of publication	Agriculture, hunting, forestry and fishing (2)	Extractive and utility supply (3)	Manufacturing	Construction	Service	All industries
1987 (4) (1)	98.5	132.10	157.1	180.59	53.0	90.33
1988	109.2	104.63	147.9	185.58	50.5	85.21
1989	109.7	93.12	151.9	183.95	47.9	83.10
1990	96.5	83.14	153.4	204.59	49.5	84.75
1991	102.9	79.58	149.3	209.23	51.8	84.83
1992 (1)	96.8	83.20	131.9	198.00	47.0	76.67
1993	118.7	73.10	123.9	174.03	49.7	75.23
1994 (6)	129.6	204.90	114.8	149.81	49.2	74.52
1995	116.3	148.15	117.9	154.84	50.5	75.01
1996	111.2	178.71	111.0	146.88	46.9	71.53
1997 (5)	191.4	249.58	174.3	236.25	85.5	116.14
1998	191.2	250.22	187.1	243.48	83.6	117.29
1999	186.6	199.46	175.8	265.06	79.1	111.92
2000	205.5	181.32	178.3	260.59	79.1	111.51
2001	195.3	157.37	176.1	244.89	74.5	105.25
2002	207.2	153.03	169.5	239.01	80.2	108.19
2003	204.4	154.50	167.7	242.29	81.2	107.98
2004 (1)	177.3	153.12	170.0	226.14	94.8	116.77
2005	166.8	147.65	169.3	205.51	94.6	114.70
2006	153.3	150.10	153.4	202.42	89.8	108.34
2007	142.8	126.41	145.8	195.65	89.5	106.41

Source HSE. The year shown correspond to the year these figures were published (i.e. 2007 = FY2006/07) (1) Standard industrial classification of economic activities used SIC80 for 1987–1991, SIC92 for 1992–2003, SIC03 for 2004–2008. (2) Excludes sea fishing. (3) Includes the number of injuries in the offshore oil and gas industry collected under the offshore installations safety legislation until 1997. These started to be included into the main statistics from 1997. (4) 1987–1996 reported under RIDDOR 1985. (5) 1996/97 onwards reported under the RIDDOR 1995. Please note that a wider number of injuries became reportable from 1996 and the HSE argues that data for non-fatal injuries collected for RIDDOR 1985 and 1996 are not comparable. (6) In the year between 1993 and 1994 the extractive and utility supply industry experienced a sharp increase, it is unknown whether this sharp difference is the real trend or correspond to a change to the recording system. No changes have been recorded in the original data source. This incongruence also appears on the over-3-days table and graph was the **manufacturing**, where incidents fell by 21.1 and 8.8% respectively in the 1987–1996 and 1997–2007 periods. The **construction** industry also showed improvements from the 1980s. Major non-fatal incidents fell by 14.6 and 12.6% respectively in the 1987–1996 and 1997–2007 periods. The **agriculture, hunting forestry and fishing** sector witnessed an increase of 10.8% in the period 1987–1996, followed by a fall of 9.6% in the period 1997–2007. The services sector also recorded an increment in major non-fatal incidents between the periods 1987–1996 and 1997–2007. These fell by 3.7% in the 1987–1996 period, but increased by 9.9% in the 1997–2007 period. The **extractive and utility supply industries** sector recorded the most significant fall in trends of major non-fatal incidents, but that only occurred recently. In the 1997–2007 period, non-fatal major incidents fell by 29.0%. That was the fastest improvement recorded in non-fatal major incidents in that period. However, the incident trends for the period 1987–1996 were also the worst in the economy, when non-fatal major incident increase by 39.7% (Chart 2).

Over-three-day injuries[5] can also be analysed (see Table 4).[6] Overall, the over-three-day injuries rate fell steadily by 11.2% in the periods 1987–1991 and 1992–1996 and by 10.7% in the periods 1997–2001 and 2002–2007. Once again, the sector that improved the most was **construction**, where over-three-day incidents fell by 23 and 23.9% in the periods 1987–1996 and 1997–2007 respectively. The second best was the **extractive and utility supply industry**, where incidents fell by 4.4 and 40.9% respectively. The **manufacturing industry** was the third best, where over-three-day injuries fell by 12.1 and 14.5% respectively. The **service** sector did not improve significantly: over-three-day injuries fell 0.3 and 3.6% in the periods 1987–1996 and 1997–2007

---

[5]Injuries resulting in more than three days off work, excluding fatal injuries.

[6]In order to make meaningful and accurate analysis of occupational health and safety incident trends since the end of the 1980s, the analysis conducted for fatal injuries in Britain has been done through the use of five- and six-year averages. Although this system might not offer a very accurate representation of any improvements occurring in recent years, it represents a useful tool for analysing longer trends, and helps reduce the errors and misinterpretations caused by sharp changes of incidents between years. The periods used for fatal injuries only are 1986/87–1990/91 (1987–1991), 1991/92–1997/98 (1992–1998), 1998/99–2002/03 (1999–2003), and 2003/04–2007/08 (2004–2008).

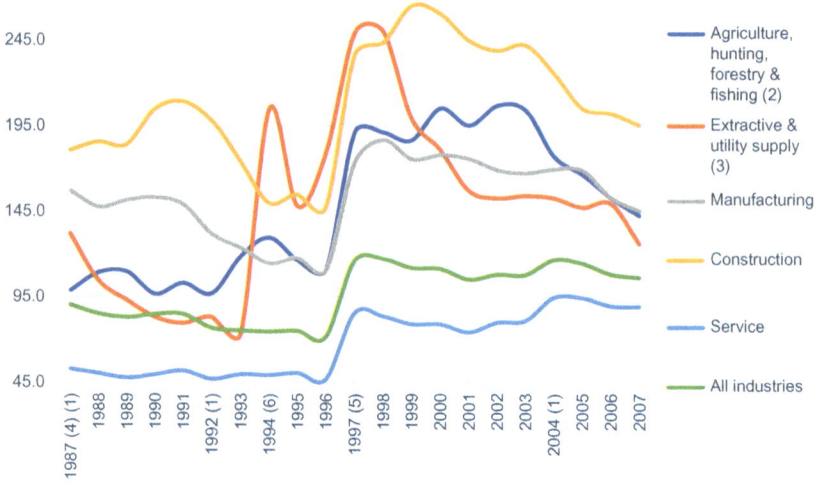

Chart 2 British non-fatal major occupational safety injuries per 100,000 workers (see Table 3 for explanatory footnotes)

respectively. The only sector that on average recorded a rise in incident trends since the 1980s is the **agriculture, hunting forestry and fishing** sector. Over-three-day injuries increased 16.8% in the period 1987–1996, and fell by 13.7% in the period 1997–2001.

Overall, it can be argued that there has been a general decline in OHS related injuries from the 1980s to the 2000s. Incidents trends fell sharply up to the end of the 1990s, but these improvements have since slowed down (Chart 3).

1.2   British Enforcements' Trends

The statistics referring to the enforcement activities of the HSE, local authorities (LAs) and the Office of Rail Regulation (ORR) are reported through HSE reports, but the definitions used between years changes and creating a longitudinal analysis is, at best, challenging. This has been a long-standing issue (Tombs 2003), and in 2013 the UK Statistics Authority also mentioned that the HSE annual statistics "tends to include little description about trends over time or contextual information about the factors influencing the statistics" (UK Statistics Authority 2013, p. 2).

Table 4 British over-three-day occupational safety injuries per 100,000 workers

Year of publication	Agriculture, hunting, forestry and fishing (2)	Extractive and utility supply (3)	Manufacturing	Construction	Service	All industries
1987 (4) (1)	226.3	1504.96	1139.27	975.50	430.42	676.0
1988	271.8	1178.94	1066.23	969.45	434.79	655.7
1989	303.9	1009.71	1137.14	882.84	431.96	648.5
1990	310.8	849.18	1231.16	920.94	443.96	652.5
1991	262.3	761.43	1219.52	994.54	453.75	651.1
1992 (1)	310.3	700.78	1175.51	970.19	450.18	632.9
1993	343.7	581.67	1087.53	806.38	444.17	597.1
1994 (6)	314.0	1512.75	947.85	644.47	433.37	569.7
1995	317.0	1189.70	992.60	658.61	445.97	581.2
1996	320.6	1084.86	889.11	574.90	413.91	536.3
1997 (5)	372.4	1092.89	838.84	563.29	416.40	513.2
1998	354.6	1221.46	880.85	577.74	429.57	530.8
1999	356.6	1068.81	838.47	545.15	423.95	512.9
2000	412.2	899.78	875.29	576.39	425.24	517.8
2001	416.9	791.84	901.63	509.54	416.24	504.0
2002	476.2	753.84	327.18	504.29	410.83	488.1
2003	420.3	659.15	320.11	491.55	406.83	477.6
2004 (1)	311.0	685.66	791.86	430.22	431.77	482.6
2005	265.5	562.22	721.03	378.83	405.14	444.7
2006	254.4	501.84	672.38	379.49	398.07	431.2
2007	252.6	434.96	616.01	347.45	390.20	414.3

Source HSE. The year shown correspond to the year these figures were published (i.e. 2007 = FY2006/07) (1) Standard Industrial Classification of Economic activities used SIC80 for 1987–1991, SIC92 for 1992–2003, SIC03 for 2004–2008. (2) Excludes sea fishing. (3) Includes the number of injuries in the offshore oil and gas industry collected under the offshore installations safety legislation until 1997. These started to be included it into the main statistics from 1997. (4) 1987–1996 reported under RIDDOR 1985. (5) 1996/97 onwards reported under the RIDDOR 1995. Please note that a wider number of injuries became reportable from 1996 and the HSE argues that data for non-fatal injuries collected for RIDDOR 1985 and 1996 are not comparable. (9) In the year between 1993 and 1994 the extractive and utility supply industry experienced a sharp increase, it is unknown whether this sharp difference is the real trend or correspond to a change to the recording system. Also, no changes have been recorded in the original data source. This incongruence also appears on the major incidents table and graph

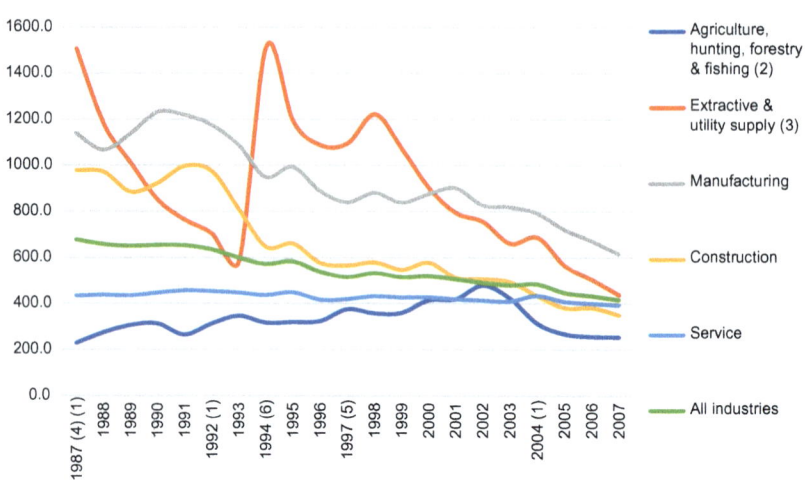

Chart 3   British over-three-day occupational safety injuries per 100,000 workers (see Table 4 for explanatory footnotes)

According to Tombs and Whyte (2010), between 2000 and 2010 the HSE decreased the number of prosecutions (0.15—0.07/100 firms) and convictions (0.115—0.06/100 firms) by half. Given the 30% budget reduction imposed by the government between 2010 and 2015, it is unlikely that the situation has improved in recent years. According to the HSE Annual statistics reports, fines issued following prosecutions on average increased from £2572 in 1995/96 to £22,321 in 2012/13 and to £126,173 in 2016/17. The sharp increase since 2013 has been caused by new sentencing guidelines (HSC 2000b, 2013e, 2017b). It is also important to consider that the Fee for Intervention (FFI) policy penalised non-compliant firms during proactive enforcement activities for a total of £36.4 million between October 2012 and April 2017.

Given that the incident trend since the end of the 1990s has slowed down, the reduction in the number of notices and prosecutions, together with a fall in the average quantity of penalties per 100 firms since 2006, means that the HSE has decriminalised OHS breaches by reducing deterrence-driven enforcement activities. Between 2000 and 2009 the number of prosecutions fell by 5% from an average of 92 to 87% of fatal injuries, and convictions by 4% from 90 to 86% of fatal injuries. This trend confirms that on average prosecution decreased faster than the number of fatal injuries from 1999 to 2009. It is also important to note that prosecution and convictions decreased after the implementation of the 2004 *Strategy for workplace health and safety to 2010 and beyond.*

This fall in enforcement activity might also be associated with the fall in numbers of front-line inspectors. Between 2002 and 2009 the number of inspectors fell by 10% from 1458 to 1323 (see various HSC and HSE Annual Reports and Accounts). This suggests that fewer firms might be inspected, but this also depends on the amount of discretion adopted by inspectors and the institutions. For example, the enforcement institution may decide to be more resource-efficient by selecting easy-to-win cases. This can be achieved by reducing the number of front-line officers, by campaigning for higher penalties and longer sentences for OHS breaches, and by offering more advice to firms.

1.3   Summary

British OHS incidents have fallen significantly since the 1980s, but the rate of improvement slowed down in the 1990s. One of the most significant issues of official OHS figures recorded in England, Wales and Scotland is that they are subject to gross mis- and under-reporting and exclude all of the work-related road incidents. RoSPA calculates that the latter cause an extra 1300 fatalities and 13,000 non-fatal injuries per year. This applies to both the official figures and the LFS. Although media attention is mostly concentrated on official figures, in the past years the HSE has also been using LFS figures to calculate the "real" social harm caused by OHS incidents. That means that incidents per 100,000 workers listed in the tables above might be as high as 730.9 for over-three-day injuries in 1986/87 and as low as 460.9 over-three-days injuries in 2007/08. These underestimates have not been taken into account in this section, but will be included at the end of the chapter when compared to the Italian figures.

The reduced improvement in incident trends at the end of the 1990s led the New Labour government to promote a strategy for revitalising OHS regulations, which focused mostly on reforming the OHS enforcement policies adopted. The innovative enforcement policies implemented by the HSE since the beginning of the 2000s, especially after the 2004 *Strategy for workplace health and safety for Britain to 2010 and beyond* and the 2012 *Löfstedt* report, have led to a decrease in proactive deterrence-driven enforcement activities, such as inspections, notices, prosecutions and convictions, and a reduction in front-line enforcement officers. Penalty levels have increased significantly since the 1990s, but a major difference occurred when the new FFI policy was enforced in 2012 and when new sentencing guidelines were introduced in 2013. The HSE commitment to deterrence-driven strategy decreased through a reduction of enforcement activities, caused by significant budget cuts, but has managed to introduce an inspection charging system (e.g. FFI) and improve the sentencing guidelines.

2   Italy

This section explores trends of OHS incidents recorded by INAIL and enforcement activities conducted by the Prevention Services for Safety in Work Environments (SPSAL) in Italy. It tries to capture the position regarding Italian incident trends and enforcement activities since 1985 and 2000 respectively.

Italian law defines a working incident as violent occurrence during working hours, which leads to the worker's death or to a temporary, permanent, absolute or partial disability, and which has caused the worker's absence from work for at least three days (Repubblica Italiana 1965 as amended and modified). The Italian official statistics also record road incidents that occur during working hours and those of workers travelling to and from work.[7] The INAIL insurance scheme does not count incidents that have happened to one-man self-employed firms or people that are not involved in any part of the working activities, such as landlords. Casual workers are also excluded, but it is not clear whether these exclude the workers that have been employed with

---

[7]In Italian these are called *itinere*, or itinerate incidents.

temporary, voucher-based and zero-hours contracts between 1999 and 2015. Another figure that INAIL excludes, when compared to British figures, is the number of injuries that have been caused to members of the public by work activities (Repubblica Italiana 1965 as amended and modified).

This section is divided into two parts. The first part is a quantitative descriptive analysis of the OHS incident trends from 1985 to 2005 by broad economic activities and type of incident. The second part explores SPSAL enforcement activity data trends in the 2000s. Labour Territorial Inspectorates (ITL) data is not available.

2.1   Italian Incident Trends

2.1.1 Italian Fatal Incidents

Overall, fatal incident trends have now decreased in Italy. However, they did increase by 6.6%, from 5.6 to 6.0 fatal incidents per 100,000 workers in the period averages 1985–1987 and 1988 1990 periods, but decreased by 34.1%, from 6.5 to 3.8 fatal incidents per 100,000 in the periods 1991–1995 and 2001–2005.[8] After an initial increase in fatal injuries in the 1980s, the 1990s were characterised by a fall of 14.1% in fatal incident trends per 100,000 workers. However, the sharper fall occurred in the period 2001–2005 when overall fatal incidents fell on average by 23.3% from the 1996–00 period (see Table 4).

The sector that recorded the best improvement was the **extractive and utility supply** industries, which recorded a 23.2% fall in the rate of fatal injuries between the mid-80s and the beginning of the 2000s. These fell by 27.5% at the end of the 1980s and by a further 29.8% between 1991 and 2005. The second best-improving sector was **construction**, which recorded a 9.9% average fall in fatal injuries

---

[8]The analysis of the fatal incidents per 100,000 workers is conducted between three- and five-year periods. This is to compensate for annual disparities and to observe lengthier trends. Therefore, the three and five-years periods are 1985–1987, 1988–1990, 1991–1995, 1996–2000 and 2001–2005. In addition, the comparison of the figures between 1985 and 1990 and those between 1991 and 2005 are not fully comparable due to the industrial classification changes from ISIC-2 to ISIC-3.

between the 1985–1987 and 2001–2005 periods. While the industry recorded one of the worst increases in fatal incidents in the economy in the 1985–1987 period (+15.6%), between 1991 and 2005 these fell by 35.3%, which represents a record improvement. The third best improvement was recorded in the **agriculture, hunting and forestry** industries, where fatal injuries fell on average by 8.3% from the periods 1985–1987 to 2001–2005. After an increase of 13.2% in fatal injuries in the second half of the 1980s, between 1991 and 2005 these fell by 29.8%. The **service** sector recorded a 6.2% average decline in fatal injuries between the 1985–1987 and the 2001–2005 periods. After an increase of 12.8% in fatal injuries in the 1988–1990 period, between 1991 and 2005 these fell by 25.2%. The only sector that recorded a slight increase in fatal injuries is **manufacturing**, which recorded a 0.5% average increase rate of fatal injuries between the 1985–1987 and the 2001–2005 periods. After an increase of 27.7% in fatal injuries in the 1988–1990 period, between 1991 and 2005 these fell by 26.7% (Table 5 and Chart 4).

2.1.2 Italian Non-fatal Incidents

Over-three days injuries in Italy (see Table 6)[9] also witnessed a fall of 21.4% from a three years average of 2980 per 100,000 workers in the years 1994–1996 to 2350 in 2003–2005. The sector that improved the most was the **manufacturing** industry, where over-three-day incidents fell by 26.8% between 1994 and 2005. The sharpest improvement was recorded in the 2003–2005 period when trends improved by 14% from the 2000–2002 period average. The second-best sector was the **agriculture, hunting forestry and fishing** sector, where over-three-day injuries decreased by 25% in the period 1994–2005. However, while the rate of over-three-day injuries fell by 24% between 1994 and 2002, in

---

[9]The analysis of the non-fatal incidents per 100,000 workers is conducted between three-year periods. This is to compensate for annual disparities and to observe lengthier trends. The three years periods are 1994–1996, 1997–1999, 2000–2002 and 2003–2005.

Table 5 Italian rate of fatal occupational safety incidents per 100,000 workers

Year of publication	Agriculture, hunting, forestry and fishing	Construction	Extractive and utility supply	Manufacturing	Service	All
1985 (1) (3)	22.18	19.90	30.87	4.05	2.17	6.23
1986	20.03	18.74	25.45	2.84	1.90	5.32
1987	20.33	18.28	26.20	3.46	1.71	5.27
1988	22.60	22.64	29.34	3.41	2.19	5.90
1989	25.90	20.59	13.00	4.86	2.24	6.11
1990	22.31	22.56	17.46	4.95	2.10	5.84
1991 (2)	21.98	26.50	8.45	6.13	2.90	6.69
1992	22.73	24.92	7.38	6.07	2.93	6.61
1993	22.23	20.32	18.64	5.86	2.74	6.13
1994	13.97	20.61	15.19	5.13	2.47	5.18
1995	13.83	19.79	10.59	5.37	2.55	5.13
1996	14.62	18.31	11.70	5.53	2.27	4.88
1997	14.08	18.10	12.13	5.86	2.77	5.21
1998	12.51	19.32	14.42	6.13	3.02	5.41
1999	14.28	18.44	7.41	5.42	3.02	5.19
2000	14.78	17.18	7.03	5.12	2.77	4.87
2001	12.52	16.43	10.01	4.73	2.44	4.45
2002	13.15	13.55	8.22	3.85	2.00	3.77
2003	11.89	15.95	8.52	4.55	1.89	3.93
2004	15.67	13.60	9.35	4.07	1.80	3.76
2005	13.22	13.00	6.42	3.74	2.03	3.69

Source INAIL (incidents) and Italian LFS (workers). (1) Industrial classification ISIC-Rev 2 used from 1985 to 1990 included. (2) Industrial classification ISIC-Rev 3 used from 1991 to 2005 included. Comparison between these two periods should be cautious. (3) All figures include work-related road incidents

the period 2003–2005 these only fell by 1%. The third best improvement in over-three day injuries was recorded in the **construction** industry, where these fell by 22.6% between 1994 and 2005. Next was the **extractive and utility supply** industry, where over-three-day injuries fell by 7.3%. While this industry experienced a 4.8% rise in over-three-day incidents between the periods 1994–1996 and 2000–2002, the period 2003–2005 recorded a 13.4% fall. The **service** sector is the only sector that experienced rising numbers of over-three-day incidents

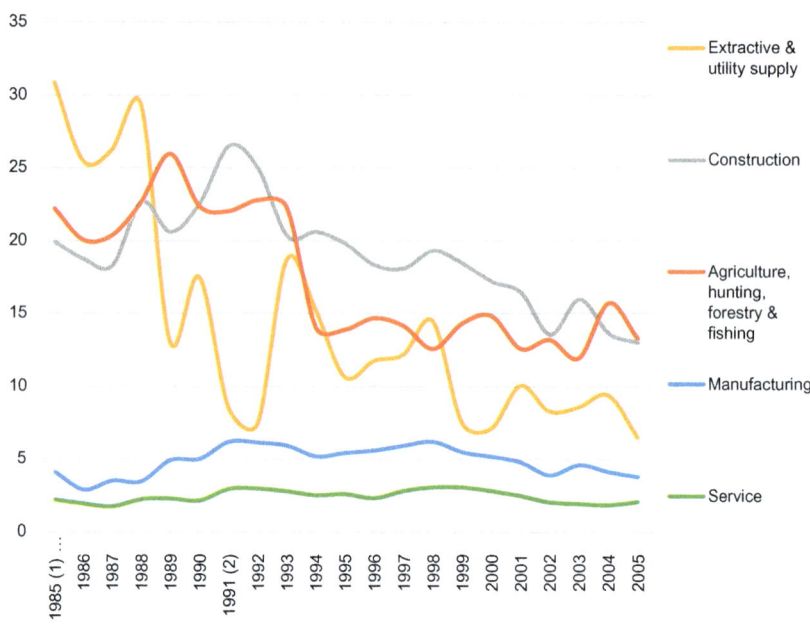

Chart 4 Italian rate of fatal occupational safety incidents per 100,000 workers (see Table 5 explanatory notes)

by 0.5% between 1994 and 2005. However, while incidents increased by 7.7% from 1994 to 2002, the sector recorded a fall of 7.2% in the three-year period 2003–2005 (Chart 5).

2.2   Italian Enforcement Trends

SPSAL enforcement activities trends are available from 2007 to 2011. There are also some regional figures available since 2000, but these are scattered and not representative of national trends. From the data available it can be observed that the number of inspectors and inspections increased in the 2000s. Since the beginning of the 2000s SPSAL increased the use of deterrence-driven enforcement policies, while they spent fewer resources on compliance-driven ones. This section explores and analyses the national enforcement trends between the years 2007 and 2011, and some regional trends from the year 2000.

Table 6 Italian rate of over-three-day injuries per 100,000 workers

	Agriculture, hunting, forestry and fishing	Construction	Extractive and utility supply	Manufacturing	Service	All
1994 (1) (2)	8193	6836	3753	5028	1517	3141
1995	7779	6338	3406	4961	1412	2972
1996	7633	6060	3328	4756	1405	2852
1997	7177	5693	3337	4528	1386	2728
1998	6890	5740	3375	4563	1453	2749
1999	6795	5957	3914	4533	1586	2816
2000	6449	5786	4108	4548	1619	2793
2001	5867	5258	3709	4360	1552	2636
2002	5613	5101	3315	3946	1496	2473
2003	5944	5149	3232	3748	1473	2416
2004	5857	5032	3233	3587	1473	2367
2005	5895	4700	3257	3458	1410	2267

Source INAIL (incidents) and Italian LFS (workers). (1) Trends have been created by summing non-fatal incidents that have caused a permanent invalidity and over-3-days one. (2) Work-related road incidents included

The number of total enforcement activities conducted by SPSAL between 2006 and 2011 increased by 30.47% from 94,533 inspections in 2008 to 123,341 in 2011 (see Table 7). The sector that recorded the largest increase in inspections is the construction sector, where these rose by 13,226 or 31%, from 41,457 in 2007 to 54,683 in 2011. The highest percentage increase occurred in the agricultural sector, where between 2007 and 2011 enforcement activities increased by 105.8% or by 3915 visits. Asbestos-related visits also increased by 22.45% or 2345 inspections from 10,459 in 2007 to 12,807 in 2011. However, the enforcement activities for industrial hygiene fell by 52.7% or 1680 visits. The national average inspection rate per firms in the territory rose by 2.13% from 4.48% in 2006 to 6.61% in 2011. Also, the number of regions meeting the 5% enforcement target set in the Health Pact rose only from 13 in 2008 to 15 regions out of 21 in 2011.

While the use of deterrence-driven enforcement activities increased from 2006, since at least 2008 SPSAL have decreased the number of

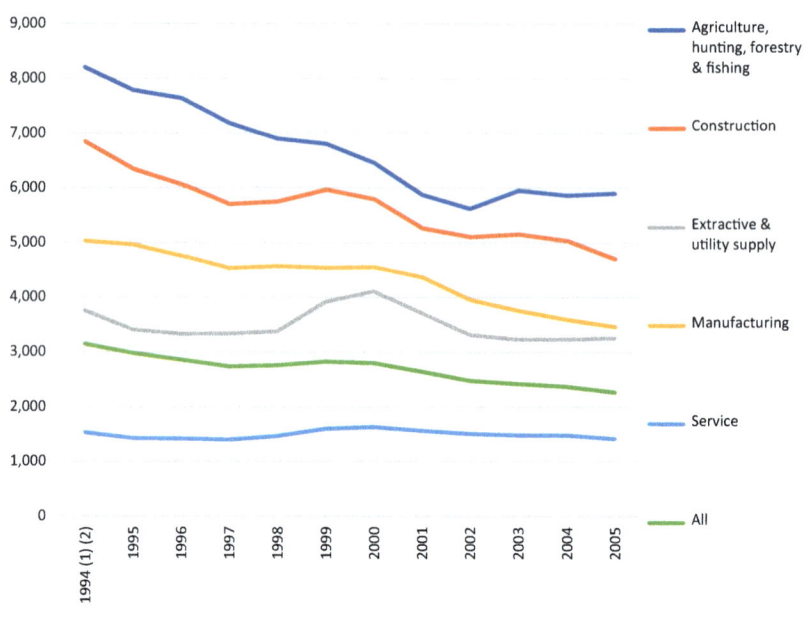

Chart 5 Italian rate of over-three day's injuries per 100,000 workers (see Table 6 for explanatory notes)

compliance-driven enforcement activities. The responsibility of conducting compliance-driven policies, through the provision of grants, information advice and training has been transferred to INAIL. Between 2002 and 2010 INAIL made available a total of over €200 million in grants for SMEs and between 2011 and 2007 this has risen to a total of €1.8 billion (INAIL 2003, 2007, 2014, 2016, 2017). Hence, between 2009 and 2011 SPSALs medical visits fell by 5.65% or 2798. The number of hours spent training stakeholders and the number of stakeholders trained by SPSAL officers also fell by respectively 0.6 and 2.9% (see Table 8). Between 2008 and 2011 SPSAL construction sites recommendation visits for buildings designed to be used for working activities also fell by 14.4% or 6351 visits. As consequence of this reduction in compliance-driven policies, between 2009 and 2011 the number of medical complaints reported by independent OHS medics increased by 53% or 1148. Although the training offered by SPSAL's

Table 7 SPSAL deterrence-driven enforcement activities 2006–2011

	2006	2007	2008	2009	2010	2011	Difference between first year available and 2011	
							Actual number	%
Total number firms visited as part of annual planned enforcement activities			94,533	119,460	119,994	123,341	28,808	30.47
Construction sites inspected		41,457	51,913	54,343	53,165	54,683	13,226	31.90
Agricultural firms inspected		3701	4178	4740	5980	7616	3915	105.78
Firms inspected for asbestos			10,459	11,926	11,999	12,807	2348	22.45
Firms or construction sites inspected for industrial hygiene regulations		3552	3658	2261	3519	1872	−1680	−52.70
Percentage of firms in the territory inspected (%)	4.48	5.07	5.37	6.78	6.63	6.61	2.13	47.54
No. of regions that have reached the 5% enforcement target (n. 21)			13	14	13	15	2	9.5

Source Coordinamento Tecnico Interregionale, elaborazione PREO as reported by Conferenza delle regioni e delle provincie autonome (2011)

staff decreased only by a small percentage, it can be argued that policy makers are trying to create an environment where information, education and training for OHS is funded by INAIL and provided by universities, the private sector, such as consulting services, and safety managers and workers' representatives.

The overall number of breaches detected by SPSAL enforcement officers between 2009 and 2011 increased by 4273 or 7.9% (see Table 8). However, it is interesting to note that an increase in the number of inspections also resulted in an increase of 7.9% to the ratio of detections per number of firms visited. This means that more resources spent on enforcement activities might increase the level of detection of incompliant firms, increase the amount of deterrence for OHS crimes, increase regulatory compliance, and decrease the number of incidents. Nevertheless, these trends also depend on the enforcement targeting strategies adopted by the local SPSALs institutions. It is also possible to note that an increase in enforcement visits of 31.9% in the construction industry did not cause an increase in number of breaches detected, but a fall of 14.5% from the figures recorded in 2007. These trends show that an increase in enforcement activities might also cause the construction sector to respond to the threat of penalties and prosecution by complying with regulations, which caused a fall in the number of breaches detected. These figures demonstrate that in the construction sector the use of deterrence-driven enforcement policies work. However, in the agricultural sector the outcome was the opposite, which means that despite an increase in enforcement of 105.8% between 2007 and 2011, the number of firms found in breach of OHS legislation increased by 139.8% between 2008 and 2011. An increase in number of inspections led to a 34% net increase in detections. The ratio of number of firms detected to the number of firms visited increased by 5% from 15.9% in 2008 to 20.9% in 2011. It might be argued that in this case the agricultural sector did not respond to the threats of detection caused by an increase in SPSAL enforcement activities, or that regulatory non-compliance was more widely spread than in the construction sector (Table 9).

The number of breaches detected following safety incident investigations and occupational health-related investigations show that between

Table 8 SPSAL compliance-driven enforcement activities 2008–2011

	2008	2009	2010	2011	Difference between first year available and 2011	% difference between first year available and 2011 (%)
Number of medical visits conducted by SPSAL (including complaints from private health and safety medics)		49,546	45,148	46,748	−2798	−5.65
Number of medical complaint that SPSAL received from private health and safety medics (included in previous line)		2142	2'95	3290	1148	53.59
Total number of inspections conducted to express opinions on constructions plans of new working sites	43,977	39,203	42,531	37,626	−6351	−14.44
Number of hours spent by SPSAL staff training stakeholders		35,733	40,229	35,510	−223	−0.62
Number of stakeholders trained by SPSAL staff		88,812	88,571	86,238	−2574	−2.90

Source Coordinamento Tecnico Interregionale, elaborazione PREO as reported by Conferenza delle regioni e delle provincie autonome (2011)

2007 and 2011 SPSALs prioritised detecting occupational health breaches rather than safety ones (see Table 10). Three observations can be made. Firstly, between 2007 and 2011 incident investigations conducted fell by 21.39%, but despite this, the legal breaches detected increased by 8.4%.[10] This is also confirmed by the fact that in the same period non-compliance increased from 29 to 36% per incident investigation conducted. Given the fall of incidents just explored, these investigation trends indicate that the enforcement institutions tried to increase the financial efficiency of the reactive enforcement activities by increasing detectable breaches, which caused an increase in number and amount of penalties collected. This technique also caused increased criminalisation of OHS breaches and more compliance, and has been followed by falling numbers of incidents and investigations. Secondly, a 15.2% rise in medical investigations between 2007 and 2011 led to a fall of 53.7% of occupational health investigations that led to the detection of legal breaches. Thus, an increase in enforcement actions conducted by SPSAL medics led to an increase in compliance. Thirdly, it is also possible to observe that the funds collected from administrative penalties fell by 18.7%, from €66.226 million in 2009 to €53.836 million in 2011. This trend might have been caused by the implementation of Law 166 2009, which decreased the level of OHS-related administrative penalties.

In addition to the national figures above, it is still possible to obtain data from some regional inspectorates and also for early years, but these trends cannot be considered representative of national enforcement trends. For example, in the region Friuli-Venezia Giulia the number of inspections conducted by SPSAL officers between 2001 and 2007 increased from 2109 to 3251. This consists respectively of 1.5 and 3.9% of the total firms present on the territory (Regione Friuli-Venezia Giulia 2010).[11] According to national data, these figures rose to 6.2%

---

[10]Figures not available from 2007.

[11]These figures represent the enforcement percentage of the total number of firms present in the region according to the Information Flows (Flussi Informativi) database. It does not represent the annual enforcement plans set by the Health Pact, which require SPSALs to conduct enough enforcement activities necessary to ensure that at least 5% of the workers in the region are working in adequate conditions.

Table 9   SPSAL proactive enforcement actions started 2007–2011

	2007	2008	2009	2010	2011	Difference between first year available and 2011	% difference between first year available and 2011 (%)
Total number firms visited as part of annual planned enforcement activities		94,533	119,460	119,994	123,341	28,808	30.47
Total number of firms detected in breach of regulations			53,895	53,939	58,168	4273	7.93
% of firms detected in breach of regulations per total firm visited			45.12%	44.95%	47.16%	2	4.52
Construction sites inspected	41,457	51,913	54,343	53,165	54,683	13,226	31.90
Construction sites found incompliant	21,682	22,999	21,546	19,443	18,530	−3152	−14.54
Percentage of incompliant construction firms detected	52.3%	44.3%	39.6%	36.6%	33.9%	−18.4	−35.18
Agricultural firms inspected	3701	4178	4740	5980	7616	3915	105.78
Agricultural firms found incompliant		663	763	1055	1590	927	139.82
Percentage of incompliant agricultural firms detected		15.9%	16.1%	17.6%	20.9%	5	31.44

Source Coordinamento Tecnico Interregionale, elaborazione PREO as reported by Conferenza delle regioni e delle provincie autonome (2011)

Table 10 SPSAL reactive enforcement actions 2007–2011

	2007	2008	2009	2010	2011	Difference between first year available and 2011	% difference between first year available and 2011 (%)
Incident investigations conducted	21,573	21,682	19,273	16,337	16,958	−4615	−21.39 (−12% from 2009)
Incident investigated where legal breach has been detected			5624	5241	6097	473	8.41
% of incident investigations where legal breaches have been detected			29%	32%	36%	7	24.13
Occupational health sicknesses investigations conducted	8603	10,417	10,214	8863	9909	1306	15.18 (−2.98% from 2009)
Occupational health sicknesses investigations caused by non-compliance			1948	1076	902	−1046	−53.70
% of occupational health sicknesses investigations where legal breaches have been detected			19%	12%	9%	−10	−52.63
Total import of all administrative penalties collected (millions of Euros)			66,226	55,564	53,836	−12,390	−18.70

Source Coordinamento Tecnico Interregionale, elaborazione PREO as reported by Conferenza delle regioni e delle provincie autonome (2011)

of firms by 2009 and fell back to 5.5% in 2011. The number of investigations for safety incidents rose from 546 in 2000 to 780 investigations in 2010. The number of occupation-related diseases also increased from 500 in 2000 to 1213 in 2010. In the same period the number of staff employed by SPSAL in Friuli-Venezia Giulia increased from 72 in 2001 to 95 in 2010, and the number of judiciary police officials (UPG)—front-line officers—increased steadily from 28 in 2001 to 68 in 2010. The number of breaches detected increased from 771 in 2000, to 1485 in 2009, but these fell back to 704 in 2010. SPSALs in Friuli-Venezia Giulia also increased the working hours dedicated to compliance-driven enforcement activities. The hours employed to support firms by SPSAL increased from 7220 in 2000 to 10,186 in 2007 (Regione Friuli-Venezia Giulia 2010). A similar trend also appeared in the statistics collected by Fooks et al. (2007), which showed that in the region of Piemonte the number of total investigations following incidents increased from 5404 in 2001 to 6386 in 2004. Finally, the number of breaches detected also increased from 8239 in 2001 to 8313 in 2004 (Fooks et al. 2007, p. 484).

2.3 Summary

In terms of safety incident trends and enforcement activities, several conclusions can be made. Firstly, the Italian trends show that ratio of incidents has decreased since the 1980s. However, overall the improvement has been more rapid since 1994. In other words, the Italian OHS regulatory and enforcement policies reforms since 1994 are likely to have caused a faster improvement in incidents trends. This improvement has been caused by the adoption of the principles-based regulation, the SPSALs use of deterrence-driven enforcement policies and a growing commitment of compliance-based strategies through the distribution of grants for small businesses by INAIL.

In terms of enforcement activities, it can be concluded that in Italy between 2009 and 2011 OHS deterrence-driven enforcement activities increased, while the penalties issued decreased. The lack of investment in reactive enforcement activities (i.e. incident and health sicknesses investigations) also fell, but the legal breaches detected per investigation conducted have increased by 7%. Given that there have not been changes to the legislation, the rise in the ratio of detection per investigation conducted demonstrates that SPSALs might have been instructed to report and penalise employers for more breaches than they used to in the past. This might be an attempt by the public prosecutors within the public ministry (PM) to increase the level of control.

Since 2000, Italian governments have passed laws designed to increase investment in OHS policies and enforcement activities. INAIL grants, designed to help small, medium and micro-size enterprises, have witnessed a significant increase. The increase in deterrence-driven enforcement activities and funding to SPSAL, as well as increased compliance-driven activities witnessed since 2000 for some regions and from 2007 nationally, has been followed by a constant fall in incident trends. Since 1994 the ratio of deterrence-driven proactive enforcement activities conducted by SPSALs has remained constant at 5% of firms on the territory. The trends of the OHS Italian policies shown in this analysis demonstrate that Italy has invested in compliance-driven OHS enforcement policies, but, unlike in Britain, also preserved a meaningful commitment to deterrence-driven enforcement policies. Thus, despite the strict due-process practices characterising the Italian criminal justice system, OHS incident prevention remains a priority, which is achieved by a combination of SPSAL's deterrenc-driven policies and INAIL's compliance-driven ones.

3   British and Italian Statistics Compared

In both Britain, and Italy incident levels have fallen steadily in the past decades. However, the fastest improvements in incidents trends in the two jurisdictions have occurred at different times. While in the former most of the improvement has occurred up to the end of the 1990s and slowed down after that, in the latter incident trends have improved slowly until the mid-1990s and accelerated after that. After the figures for British fatal incidents are adjusted to include work-related road incidents, and the non-fatal ones to include the under- and misreported data as estimated by Tombs and Whyte (2008), the British fatal and non-fatal incidents per 100,000 workers appear higher than in Italy (see Tables 11 and 12). Due to the Italian stakeholders' incentives to report OHS incidents, Italian OHS incident figures might not be subject to the same level of under-reporting as the British figures. This is a very important observation because the claim that until the 2005 European Union (EU) enlargement Italy held the worst OHS records might not be accurate, and this might still be the case in the present day when compared to other countries. The level of mis- and under-reporting in Britain has also been used to fabricate political rhetoric undermining the real social costs of OHS incidents in Britain, which has helped justify the erosion of enforcement policies, particularly deterrence-driven ones, and further under-criminalised these crimes (Chart 6).

These trends become notable also when looking at the OHS enforcement policy reforms implemented in Britain and Italy. In Britain, the gradual diversion of enforcement funding from deterrence-driven to compliance-driven enforcement policies since the early 2000s does not appear to have managed to decrease OHS incident trends significantly. In Italy, the opposite happened. A sharper fall of OHS incidents followed from the mid-1990s regulatory and enforcement policy reforms, which increased the emphasis given to deterrence-driven policies by SPSAL. In other words, this comparative analysis suggests that principles-based regulation, enforced by deterrence-driven enforcement policies, are more effective in preventive incidents and achieving compliance, which also reduces the under-criminalisation of OHS crimes.

These statistical trends are very important for this research because the book relies on these analyses to draw more in-depth conclusions on the regulations and specific enforcement policies adopted by the OHS inspectorates in these two jurisdictions. Understanding the limitations of the OHS statistics available and the methodologies adopted to collect this data, is vital to understand the perceived gravity of these phenomena in jurisdictions and to explain the political rhetoric driving policy change (Chart 7).

Table 11   British and Italian fatal occupational safety injuries per 100,000 workers

Year of publica-tion	Italian occupational fatal incidents in all industries including work-related road incidents	British occupational fatal incidents in all industries	British occupational fatal incidents in all industries including 1300 estimated work-related road incidents (7)
1985 (1) (3)	6.23		
1986	5.32		
1987 (5) (4)	5.27	1.72	7.21
1988	5.9	1.81	7.11
1989	6.11	2.40	7.52
1990	5.84	1.85	6.93
1991 (2)	6.69	1.73	6.93
1992 (6)	6.61	1.51	6.84
1993	6.13	1.41	6.83
1994	5.18	1.23	6.62
1995	5.13	1.11	6.42
1996	4.88	1.04	6.28
1997 (4)	5.21	1.14	6.29
1998	5.41	1.07	6.15
1999	5.19	0.97	5.98
2000	4.87	0.84	5.78
2001	4.45	1.09	5.95
2002	3.77	0.94	5.8
2003	3.93	0.84	5.65
2004 (4)	3.76	0.86	5.61
2005	3.69	0.81	5.51
2006		0.78	5.44
2007		0.88	5.54
2008		0.82	5.4

Italian source INAIL (incidents) and Italian LFS (workers number). *British initial data source* HSE and UK LFS (workers number). The year shown correspond to the year these figures were published (i.e. 2007 = FY2006/07) (1) Industrial classification ISIC-Rev 2 used from 1985 to 1990 included. (2) Industrial classification ISIC-Rev 3 used from 1991 to 2005 included. Comparison between these two periods should be cautious. (3) All figures include work-re-lated road incidents, but exclude self-employed workers' one. (4) Standard Industrial Classification of Economic activities used SIC80 for 1987–1991, SIC92 for 1992–2003, SIC03 for 2004–2008. (5) 1987–1996 reported under the RIDDOR 1985. (6) 1997–2008 reported under RIDDOR 1995. (7) The estimated figures are provided by RoSPA for the year 2012/13. These figures should be interpreted with caution because work-related road incidents might have increased or decreased since the 1980s. However, these are the closest esti-mates to use for comparing British and Italian figures

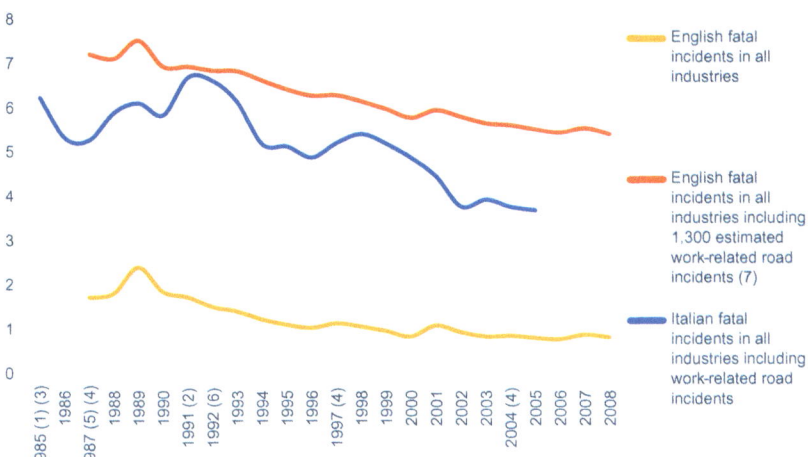

Chart 6   British and Italian fatal occupational safety injuries per 100,000 workers (see Table 11 for explanatory footnotes)

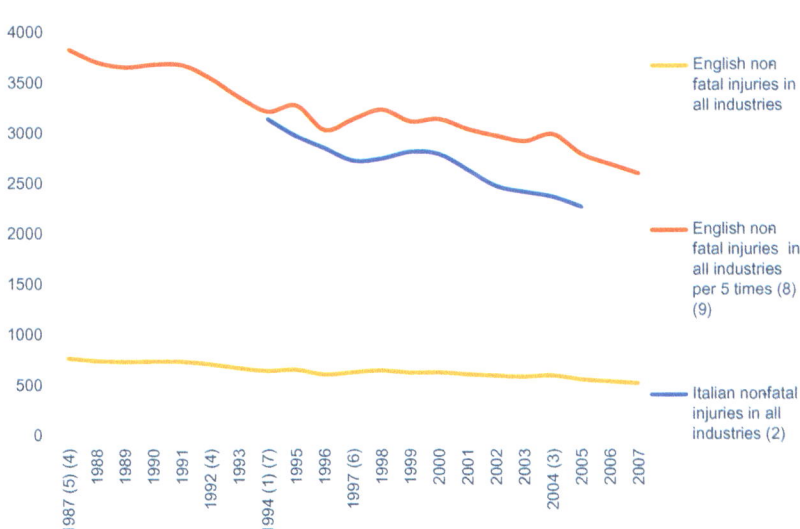

Chart 7   British and Italian non-fatal occupational safety incidents per 100,000 workers (see Table 12 for explanatory footnotes)

Table 12 British and Italian non-fatal occupational safety incidents per 100,000 workers

Year of publication	Italian non-fatal injuries in all industries (2)	British non-fatal injuries in all industries	British non-fatal injuries in all industries per 5 times (8) (9)
1987 (5) (4)		766	3832
1988		741	3705
1989		732	3658
1990		737	3686
1991		736	3680
1992 (4)		710	3548
1993		672	3362
1994 (1) (7)	3141	644	3221
1995	2972	656	3281
1996	2852	608	3039
1997 (6)	2728	629	3147
1998	2749	648	3240
1999	2816	625	3124
2000	2793	629	3147
2001	2636	609	3046
2002	2473	596	2981
2003	2416	586	2928
2004 (3)	2367	599	2997
2005	2267	559	2797
2006		540	2698
2007		521	2604

Italian source INAIL (incidents) and Italian LFS (workers number). *British initial data source* HSE and UK LFS (workers number). (1) Trends have been created by summing non-fatal and over-3-days incidents. (2) Work-related road incidents included, but excludes injuries to self-employed. (3) Standard Industrial Classification of Economic activities used SIC80 for 1987–1991, SIC92 for 1992–2003, and SIC03 for 2004–2008. (4) Includes the number of injuries in the offshore oil and gas industry collected under the offshore installations safety legislation until 1997. These started to be included into the main statistics from 1997. (5) 1987–1996 reported under RIDDOR 1985. (6) 1996/97 onwards reported under the RIDDOR 1995. Please note that a wider number of injuries became reportable from 1996 and the HSE argues that data for non-fatal injuries collected for RIDDOR 1985 and 1996 are not compara-ble. (7) In the year between 1993 and 1994 the Extractive and utility supply indus-tries experienced a sharp increase, it is unknown whether this sharp difference is the real trend or corresponds to a change to the recording system. No changes have been recorded in the original data source. This incongruence also appears on the over-3-days table and graph. (8) "[I]n order to reconstruct the data on injuries use-fully, then, we would need to apply a multiplier of much greater than five to six times HSE headline figure" (Tombs and Whyte 2008, p. 3). Hence these figures are increased by five times to compensate for the under and misreporting. (9) Excludes injuries caused by road incidents, but includes self-employed workers' injuries

References

Bibbings, R. (2014). *Twenty four arguments for increased action by the Health and Safety Commission and Executive.* Royal Society for the Prevention of Accidents [Online]. Available from: http://www.rospa.com/drivertraining/morr/info/24arguments.pdf. Accessed 25 Aug 2014.

Centre for Occupational and Environmental Health. (2013). *The health and occupation research network* [Online]. Available from: http://www.population-health.manchester.ac.uk/epidemiology/COEH/research/thor/. Accessed 25 Aug 2014.

Cherry, N., Meyer, J. D., Adisesh, A., Brooke, R., Owen-Smith, V., Swales, C., et al. (2000). Surveillance of occupational skin disease: EPIDERM and OPRA. *The British Journal of Dermatology, 142*(6), 1128–1134.

Coordinamento Tecnico Interregionale. (2011). *Attività delle regioni e delle province autonome per la prevenzione nei luoghi di lavoro, elaborazione PREO as reported by Conferenza delle regioni e delle provincie autonome* [Online]. Available from: http://www.regione.emilia-romagna.it/sicurezza-nei-luoghi-di-lavoro/coordinamento/altre-strutture-e-documenti-di-riferimento/piani-nazionali e-regionali. Accessed 25 Aug 2014.

Fooks, G., Bergman D., & Rigby, B. (2007). *International comparison of (a) techniques used by state bodies to obtain compliance with health and safety law and accountability for administrative and criminal offences and (b) sentences for criminal offences.* Health and Safety Executive [Online]. Available from: http://www.hse.gov.uk/research/rrhtm/rr607.htm. Accessed 25 Aug 2014.

Great Britain. (2013). *Reporting of injuries, diseases and dangerous occurrences regulation 2013.* N. 1471. Norwich: The Stationary Office [Online]. Available from: http://www.legislation.gov.uk/uksi/2013/1471/contents/made. Accessed 23 Mar 2017.

HSE. (2003). *Driving at work, managing work-related road safety* [Online]. Available from: http://www.hse.gov.uk/pubns/indg382.pdf. Accessed 25 Aug 2014.

HSE. (2012a). *HSE annual report and accounts 2011/2012* [Online]. Available from: http://www.hse.gov.uk/aboutus/reports/index.htm. Accessed 25 Aug 2014.

HSE. (2012b). *Historical picture: Trends in work-related injuries and ill health since the introduction of the Health and Safety at Work Act (HSWA) 1974* [Online]. Available from: http://www.hse.gov.uk/statistics/history/historical-picture.pdf Accessed 25 Aug 2014.

HSE. (2013a). *Data source* [Online]. Available from: http://www.hse.gov.uk/ statistics/sources.htm#employment. Accessed 25 Aug 2014.

HSE. (2013b) *Enforcement management model* [Online]. Available from: http://www.hse.gov.uk/enforce/emm.pdf. Accessed 25 Aug 2014.

HSE. (2013c). *National local authority enforcement code health and safety at work England, Scotland & Wales* [Online]. Available from: http://www.hse. gov.uk/lau/national-la-code.pdf. Accessed 23 Feb 2018.

HSE. (2013d). *Reporting accidents and incidents at work: A brief guide to the reporting of injuries, diseases and dangerous occurrences regulations 2013 (RIDDOR)* [Online]. Available from: http://www.hse.gov.uk/pubns/ indg453.pdf. Accessed 23 Mar 2017.

HSE. (2013e). *Health and Safety Executive statistics 2009/10* [Online]. Available from: http://www.hse.gov.uk/statistics/overall/hssh0910.pdf?pdf=hssh0910. Accessed 10 Apr 2010.

HSE. (2017). *Health and safety at work summary statistics for Great Britain 2017* [Online]. Available from: http://www.hse.gov.uk/statistics/overall/ hssh1617.pdf. Accessed 10 Apr 2017.

HSE. (2018). *Is health surveillance required in my workplace?* [Online]. Available from: http://www.hse.gov.uk/health-surveillance/requirement/ index.htm. Accessed 23 Mar 2018.

INAIL. (2003). *Bilancio Consuntivo 2002* [Online]. Available from: https:// www.inail.it/cs/internet/docs/1_bil-cons-2002-pdf.pdf. Accessed 6 Apr 2018.

INAIL. (2007). *Bilancio Consuntivo 2006* [Online]. Available from: https:// www.inail.it/cs/internet/docs/bilancio-cons-2006-pdf.pdf. Accessed 6 Apr 2018.

INAIL. (2014). *Bilancio Consuntivo 2013* [Online]. Available from: https:// www.inail.it/cs/internet/docs/all-bilancio-consuntivo-2013.pdf. Accessed 6 Apr 2018.

INAIL. (2016). *Bilancio Consuntivo 2015* [Online]. Available from: https:// www.inail.it/cs/internet/docs/ammt-bilancio-consuntivo-2015.pdf. Accessed 6 Apr 2018.

INAIL. (2017). *Bilancio Previsione 2017* [Online]. Available from: https:// www.inail.it/cs/internet/docs/ammt-bilancio-previsione-2017.pdf. Accessed 6 Apr 2018.

Johnson, J. (2010, November). Beating fatigue. *RoSPA Occupational Safety & Health Journal, 40*(11), 27–30.

Regione Friuli-Venezia Giulia. (2010). *Attività UOPSAL, ASS Friuli Venezia Giulia 2000–2010* [Online]. Avialable from: http://www.regione.fvg.it/

rafvg/export/sites/default/RAFVG/salute-sociale/organizzazione-salute-tu-tela-sociale/FOGLIA29/allegati/AttivitaSPSAL00-10xcomart27xsito.ppt. Accessed 25 Aug 2014.

Repubblica Italiana. (1965). *Decreto Del Presidente Della Repubblica n. 1124, 30 giugno 1965.* Testo unico delle disposizioni per l'assicurazione obbligatoria contro gli infortuni sul lavoro e le malattie professionali. *Gazzetta Ufficiale* n.257 del 13 ottobre 1965, Supplemento Ordinario N. 24.

Repubblica Italiana. (2015). *Decreto Legislativo* 4 marzo 2015, n. 22 Disposizioni per il riordino della normativa in materia di ammortizzatori sociali in caso di disoccupazione involontaria e di ricollocazione dei lavoratori disoccupati, in attuazione della legge 183/2014. Pubblicato nella Gazz. Uff. 6 Marzo 2015, n. 54, S.O. [Online]. Available from: http://www.gazzettaufficiale.it/eli/id/2015/3/6/15G00036/sg. Accessed 30 Jan 2018.

Tombs, S. (2003). Accounting for safety crimes? HSE, enforcement data and the (shifting) politics of access. *Radical Statistics, 81*(Spring), 51–61.

Tombs, S., & Whyte, D. (2008). *A crisis of enforcement: The decriminalisation of death and injury at work.* London: Centre for Crime and Justice Studies, 1746–6938 [Online]. Available from: http://www.crimeandjustice.org.uk/sites/crimeandjustice.org.uk/files/crisisenforcementweb.pdf. Accessed 25 Aug 2014.

Tombs, S., & Whyte, D. (2010). *Regulatory surrender: death, injury and the non-enforcement of law.* Liverpool: Institute of Employment Rights.

UK Statistics Authority. (2013, September). *Assessment of compliance with the code of practice for official statistics. Statistics on health and safety at work (produced by the Health and Safety Executive).* Assessment Report 261 [Online]. Available from: https://www.statisticsauthority.gov.uk/archive/assessment/assessment/assessment-reports/assessment-report-261—statistics-on-health-and-safety-at-work.pdf. Accessed 10 Jan 2017.

4

Scrutinising Public Institutions

The challenges encountered during the fieldwork that has informed this comparative analysis are too important to be ignored. This short chapter, thus, is essential to explain the methods adopted to collect the data, and what happened when I approached the Health and Safety Executive (HSE) with the aim of conducting interviews with occupational health and safety (OHS) enforcement officers.

This research study analyses the British and Italian OHS enforcement officers' opinions on the methods used to ensure legal compliance and achieve the regulatory objectives. This research study considers the under-criminalisation of OHS crimes as a serious fundamental issue, which has grave negative consequences for workers, and is an issue that can potentially erode the legitimacy of the regulation and enforcement institutions and the criminal justice systems values of justice and equality.

The research study adopted a qualitative comparative research method to extract meaning and capture detailed, subjective viewpoints of enforcement officers (Silverman 2011; Snape and Spencer 2003; Flick 2009). Qualitative methods are used to formulate new research theories based on questions relative to the context studied, but also to find causal relationships between phenomena (Bernard 2013), and to empower researchers

© The Author(s) 2019   **125**
D. Canciani, *The Politics and Practice of Occupational Health
and Safety Law Enforcement*, Critical Criminological Perspectives,
https://doi.org/10.1007/978-3-319-98509-1_4

with tools to decide the parameters from which these contexts are to be studied (Silverman 2011; Flick 2009). This pluralisation has allowed researchers to express and promote researchers' standpoints through the use of qualitative methods that can sensitise audiences and offer hidden and innovative meanings to social phenomena.

Aspects that qualitative studies are lacking are the reliability, validity and objectivity of the findings. These methodological rigour standards, however, are considered inappropriate for qualitative studies. (Lincoln and Guba 1985; Guba and Lincoln 1994) seminal work on the *trustworthiness* of qualitative research studies illustrates that the rigour of qualitative studies should be assessed with different parameters. *Credibility* is defined as the capability of the research methodology to produce valid research findings. *Dependability* is defined as the capability of the research design to produce reliable data, which is achieved by leaving an audit trail that can trace the findings back to the original source. *Confirmability*, (Guba and Lincoln 1994; Lincoln and Guba 1985) argue, is the capacity to demonstrate that the research project fieldwork has occurred objectively, or at least to factor into the research method and fieldwork techniques to achieve the higher level of objectivity possible. *Transferability* is defined by (Guba and Lincoln 1994; Lincoln and Guba 1985) as the degree to which the findings can be generalised to the wider population. This depends on the number of participants interviewed, the time spent interviewing them and whether the overall expertise of the participants is representative of the population for which the data is generalised (see also Bryman 2008; Flick 2009). Trustworthiness was achieved by requesting interviewers to review their scripts, through the use of a qualitative analysis software capable of coding and tracking responses, and by factoring into the design of the methodology systems to achieve the highest level of objectivity possible. Transferability depends on the sample selected, which will be further discussed below.

The cross-national comparative nature of this study is also recognised as highly valuable research methodology in social sciences because it adds a more critical interpretation of current practices, an insight to the current policy debates in other jurisdictions and, thus, brings an innovative contribution to the literature on the subject (Nelken 2010; Dammer and Albanese 2011; Fields and Moore 2005). Comparative studies offer empirical and theoretical perspectives that are less ethno-centric and can help identifying best policy practices within and across borders (Nelken 2010; Dammer and Albanese 2011; Fields and Moore 2005; Cole et al. 1987). Dammer and Albanese (2011) argue that in comparative studies researchers can improve the level of analysis by distinguishing and comparing between basic principles traditionally underlying specific social, political and criminal justice system practices. These include aspects such as, in this case, the separation of powers between state institutions, the different designs of the criminal justice system, and their cultural and historical traditions. However, researchers conducting comparative studies must be careful when interpreting and comparing cultural and legal definitions. Meanings go beyond simple translations and are embedded in cultures and traditions, which need to be comprehended thoroughly by the researcher (Broadfoot 2000; Broadfoot et al. 1993). Generalisations of the findings beyond the jurisdictions analysed should be cautious because the analyses under-taken are very ethnic- and cultural-specific to and between the countries studied (Dammer and Albanese 2011). As in any research, biases are inevitable (Becker 1963), but in comparative studies, these might also be fuelled by nationalistic and patriotic sentiments (Dammer and Albanese 2011). Although differences between jurisdictions can present significant methodological hurdles for the comparison, Nelken (2010) argues that it is the differences that make comparative studies interest-ing, which makes them "debate[s] worth having" (Nelken 2010, p. 35).

Research studies analysing the enforcement policies of economic reg-ulators seem to be idiosyncratically linked to the methodological diffi-culties encountered when 'researching up'. Traditional research studies are concentrated on 'researching down', which concentrates on vulnera-ble social groups socially positioned on a lower social hierarchical level. Doing research is in itself a process of affirming the researcher's power over the design and pursuit of the research project (Walford 2011). Hence, when researching up, academics are faced with challenges. Gaining access is usually one of the major hurdles due to the fear of exposing participants' or organisational practices that might be criticised and undermine their legitimacy or elite positions. These difficulties can be reduced by building relationships with individuals and organisations

that are the subject of study, but this might require researchers to negotiate the terms of access or the research methodologies used to collect
the data (see Almond 2008 for example). The worst case scenario is
when researchers are forbidden, due to contractual agreements, from
including findings that the funding organisation might disapprove of.
Relevant examples on the subject of this study are Whyte (2000), who
refused research funding from the HSE and was consequently denied
further access to interviewees, or Hawkins (2002), who argues that the
ideas expressed in his research project, which was also funded by the
HSE, are his own "and not *necessarily* those of the HSE" (p. xiii, emphasis
added). In Hawkins's case, it is difficult to understand how much the
HSE was involved in drafting the book.

Accessing these organisations can be difficult, but younger researchers
are usually less threatening to elites, and interviewing retired personnel,
or employees willing to the take the risk of participating without the
consent of the organisation are alternative techniques for gathering data
(Gewirtz and Ozga 1994; Neal 1995; Westmarland 2011; Kogan 1994).
Therefore, researching elites and the powerful requires a more elabo-
rated and resilient methodological approach than is usually adopted
when researching down, or more vulnerable populations. At the same
time, however, exposing people in elite and powerful positions to criticism might cause as much personal and professional damage as those of
vulnerable interviewees, which must be a very important and constant
ethical consideration when conducting these types of research studies.
Academics conducting research projects on elite and powerful groups
need to be better informed and prepared about the possible issues that
they might encounter. This should be one of the priorities at every
stage of the study, from the design to the dissemination stages (Walford
2011; Fitz and Halpin 1994; Gewirtz and Ozga 1994; Ball 1994; Fitz
and Halpin 1994; Whyte 2000).

The research population of this study were OHS enforcement
officers, who enforce regulations in firms on a daily basis and operate
at the grassroots level. Their experiences allow a better understanding of
the forces, pressures and restrictions exercised within their own organisations, by stakeholders in the field and by wider sociopolitical contexts. They have a good understanding of the contexts causing policy

reforms. Enforcement officers visit businesses in person and they talk to employers, the workers' representatives and the workers themselves. They assess work environment OHS levels, and take decisions, often unsupervised. They are involved, to various degrees, in prosecutions processes. In Britain, officers take enforcement cases to court, while in Italy they only appear as witnesses. They might be considered, especially by stakeholders, as legal representatives of the law, whose knowledge is fundamental for creating a link between the interpretative nature of the regulations and the technical pragmatic safety issues found in workplaces. This research aimed to capture the officers' day-to-day issues, frustrations, and underlying personal interests they have when enforcing the British and Italian OHS regulations. The research assumed that field enforcement officers are knowledgeable and show a high commitment to improving OHS enforcement activities and practices, but that they also use their power to achieve their own goals.

There are two fundamental reasons why Britain and Italy have been selected for this comparative research study. Firstly, they have very similar regulatory frameworks but very different enforcement policies. The reason for this variance is mainly caused by the different legal and political traditions of these two countries. The anglophone literature on the subject is mostly concentrated on the adoption of regulatory enforcement policies implemented in jurisdictions in which criminal justice system policies originate from British common law and crime-control traditions. A comparative analysis with a jurisdiction where the criminal justice system is traditionally based on statute law due-process principles can add significant value to the debate. Secondly, the regulatory enforcement policies adopted in both countries differ fundamentally, but both experience an under-criminalisation of OHS crimes. In other words, despite the different social, political and criminal justice system policies and traditions of these two jurisdictions, regulatory crimes are subject to a conventional form of under-criminalisation when compared to similar harmful crimes. It is, thus, important to understand the reasons for this fundamental inequality. Through a Marxist theoretical framework, and guided by Pearce's (1976) seminal work, this research project aims to understand why this is the case and what causes this issue in both jurisdictions (Table 1).

Table 1 Summary of key interview information

	England	Italy
Number of interviews[a]	9	13
Number of Interviewees[a]	9	15
Enforcement institutions	HSE ($n = 5$); ORR[b] ($n= 1$) LA ($n = 3$)	SPSAL ($n = 12$); DPL ($n = 1$)
Interviewees' average years of employment	19	18
Interviews conducted between	September 2009 and October 2011	July 2009 and August 2009
Interviewee gender	2 Females, 7 Males	4 Females, 11 Males
Total interview time	13 h 17 Min	12 h 50 Min
Average time per interview	88 Min	60 Min
No. of audio recorded interviews	8	11
No. of districts covered	4	5
Interviewees' duties within enforcement institution	2 Chief Enforcement Officers 1 Retired Chief Enforcement Officer 5 Field Officers 1 Enforcement Officers Instructor	1 Unit Director 6 Chief Enforcement Officers 8 Field Officers
Transcripts checked	4 out of 9	8 out of 13

[a]In two Italian interviews the interviewees decided to participate with a colleague, which explains the reason the number of interviewees are more than that of interviews
[b]Please note that the ORR interviewee was working for the HSE for a number of decades before being transferred to the rail regulator (ORR). Although this project analyses the HSE's and LA's enforcement policies, the difficulties encountered in recruiting participants meant that the ORR candidate was a good enough match for the sample

1   The British and Italian Fieldworks

One of the major challenges of this research study was the fieldwork, which lasted from July 2009 to October 2011. While 13 Italian interviews were conducted in 2 months, between July and August 2009, it took 25 months, between January 2010 and October 2011, to interview 9 British officers. The contact details of Italian chief inspectors were easily accessible online or through the organisations' telephone switchboards. Five Prevention Services for Safety in Work Environments (SPSAL) chief inspectors were immediately interested in the study and offered full access to their offices and the enforcement officers operating under their directions. Only one out of five Labour Territorial Inspectorates (ITL) chief enforcement officers contacted agreed to participate in the study. During the interview with the ITL officer, it was revealed that the reason for the low response rate was caused by a negative media attention to the organisation and officers. The research study commitment to use anonymity and confidentiality was not enough to persuade more Italian ITL officers to take part.

The British recruitment process confirmed many of the challenges encountered by scholars 'researching up'. The HSE gatekeeper actively prevented me from interviewing OHS enforcement officers. The lack of contact details available for local HSE chief enforcement officers prevented me from recruiting participants in the same way I had done in Italy. Thus, I decided to approach the institution by calling the HSE switchboard. The phone call was passed on to the Head of the Field Operation Directorate (FOD) Planning Unit, who became my gatekeeper.[1] Upon requesting 13 participants, the gatekeeper promised to forward my invitation to enforcement officers based in England. I interviewed the first officer in January 2010 but by April 2010 the gatekeeper withdrew the availability. The excuses that caused the delay and, finally, the withdrawal, included the gatekeeper's misunderstanding of the interview information, which was provided throughout; lack of time due to important deadlines, such as the end or beginning of the financial and calendar year; and the interviews' length (one hour each) were judged too long. These justifications confirmed the difficulties experienced by previous researchers (Walford 2011; Westmarland 2011; Dantzker and Hunter 2012).

---

[1]Gatekeeper, in this instance, is not defined as a person controlling access to a vulnerable population, but as someone who had—simply—control over the access of the researchers to the field. That is because the health and safety enforcement officers are not considered vulnerable participants and because it is their responsibility to seek consent from their line manager for participating to research studies.

In September 2010, I tried to regain access to the OHS enforcement officers by sending a letter to the HSE gatekeeper with references from a senior academic colleague in the department. The gatekeeper did not reply, but in January 2011, a second HSE enforcement officer agreed to participate in the research. Initially, I thought that the gatekeeper had accepted the written letter sent in September 2010, but it was only later I discovered that the participant was following up the invitation I had sent in 2009. Hence, from January 2011, I wrongly presumed that I had the permission to continue the fieldwork, and to avoid being a burden to the gatekeeper, I decided to increase the research population to include local authority (LA) OHS enforcement officers and to change the recruitment approach by utilizing the snowballing sampling technique. I decided that I was going to use the addresses of the HSE local offices published online to send out letters inviting chief inspectors to participate in the study.

I sent open invitation letters to nine HSE offices in England. Three HSE chief enforcement officers responded to my letters (33%), but told me that they had to seek approval from the Head of the FOD Planning Unit, the gatekeeper. All the officers' requests were denied, which proved that the HSE gatekeeper had not sent the initial invitation letters to as many interview candidates as I had hoped for. Within hours, the HSE gatekeeper sent me an intimidating email asking me to disclose information about the people that I had interviewed, including their names and the information they gave me. The request fuelled an intense feeling of distress because it consisted of a direct threat to the ethical commitments promised to my interviewees. After consulting with a lawyer, it was suggested that the only reason I needed the HSE gatekeeper was to gain access to the field, not to seek approval for undertaking interviews. In other words, it was the interviewees' responsibility to ask permission and withhold sensitive information, not mine. Participants were asked to avoid disclosing information protected by the Official Secrets Act (1989) and the Data Protection Act (1998). With the awareness that I had committed no legal offence, I explained to the gatekeeper the reasons why I thought I could interview HSE enforcement officers, and apologised for the misunderstanding, but refused, with regret, to provide any information due to the research study commitment to the anonymity and confidentiality of participants. As a result, the gatekeeper sent me an email asking me to refrain from contacting other enforcement officers (personal email). After this occurrence, all of the HSE enforcement officers contacted via letter decided—regretfully—to withdraw their participation from the research study. This incident suggested that the gatekeeper might have tried to prevent me from completing the field-work successfully by only contacting two enforcement officers, who were interviewed in January 2010 and 2011.

After this incident, I had to find ways to recruit more participants and complete the fieldwork. All the participants contacted by letter refused to participate, but were happy to pass my request on to retired enforcement officers that might be willing to participate. By summer 2011, I managed to talk to three retired HSE inspectors. I also decided to approach LA enforcement officers, whose contribution to the study proved to be very valuable. The British fieldwork was completed in October 2011, 25 months after the first phone call with the gatekeeper and 21 months after the first interview.

2   Summary and Discussion

The experiences encountered during the fieldwork of this research study confirms that researching up is challenging. This experience also confirms the arguments of Walford (2011), Westmarland (2011) and Dantzker and Hunter (2012) that elite interviewees might try to stop scholars from researching (politically) sensitive topics, and forbid employees within their organisation from participating in research studies. Westmarland's (2011) technique of contacting retired enforcement officers and using a purposive snowballing technique proved to be effective. In addition, given that the HSE is a public institution, it is hard to justify a lack of access. Indeed, lack of time, which was used by the HSE gatekeeper to justify the recruitment delay, might be one of the most common justifications used to stop researchers accessing the field.

While planning research studies focusing on powerful individuals and organisations, it is important to carefully consider ethical issues that might cause unwelcome effects not only on the interviewees' lives and careers, and on the organisation scrutinised, but also on the researcher. Employees and researchers have been prosecuted in the past for releasing secret or classified information, which is a risk that should not be ignored when conducting these types of research studies. Anonymity and confidentiality are fundamental tools to use for protecting participants. They can also protect researchers from the possibility of being forced to release personal information through court orders. Participants should not be forced to sign consent forms, names should be deleted from transcripts as soon as possible and before dissemination, and no participant's account should be linked to the location of their offices or name of the organisation they belong to. This includes using communication channels that are not in control of the organisation researched, such as contacting participants through work emails. These ethical safeguards do not only protect the research participants, but has also the capacity to improve the quality of the data gathered. Anonymity and confidentiality reassure interviewees and encourage them to give franker responses. Regarding the interview information sheet, the interviewees should be asked to avoid releasing information protected by official secrets and privacy legislation or any information that is protected by patents owned by organisations. If interviewees choose to reveal any of this information, the researcher should think carefully whether and how the information can be disseminated (Bryman 2008; Flick 2009; Silverman 2011). Thus, ethical procedures should eliminate any legal evidence that might incriminate participants or researchers. Due to the lower hierarchical positioning of the researcher to that of the researched, it would be a mistake to think that ethical procedures are used to protect either the researcher or the research population. Adequate research ethics should protect everyone from adverse consequences when researching up.

References

Almond, P. (2008, June). Investigating health and safety regulation: Finding room for small-scale projects. *Journal of Law and Society*, Special Issue: Law's Reality: Case Studies in Empirical Research on Law, *35*(Issue Supplement 1), 108–125.

Ball, S. J. (1994). Political interviews and the politics of interviewing. In G. Walford (Ed.), *Researching the powerful in education*. London: UCL Press.

Becker, H. S. (1963). *Outsiders: Studies in the sociology of deviance*. New York: Free Press.

Bernard, R. H. (2013). *Social research method: Qualitative and quantitative approaches* (2nd ed.). Thousand Oaks, CA: Sage.

Broadfoot, P. (2000). *Interviewing in a crosscultural context: Some issues for comparative research*. In S. Hillyard (Series ed.). *Studies in qualitative methodology* (Vol. 6, pp. 53–65). London: Emerald Group Publishing Ltd.

Broadfoot, E. M., Osborn, M. J., Gilly, M., & Blucher, A. (1993). *Perception's of teaching: Teachers' lives in England and France*. London: Cassells.

Bryman, A. (2008). *Social research method* (3rd ed.). Oxford: Oxford University Press.

Cole, G. F., Frankowski, S., & Gertz, M. G. (1987). *Major criminal justice systems: A comparative survey*. Beverly Hills: Sage.

Dammer, H. R., & Albanese, J. S. (2011). *Comparative criminal justice*. Belmont: Thomson.

Dantzker, M. L., & Hunter, R. (2012). *Research methods for criminology and criminal justice* (3rd ed.). Sudbury: Jones & Bartlett Learning.

Fields, C. B., & Moore, R. H. (Eds.). (2005). *Comparative criminal justice*. Prospect Heights, IL: Waveland Press.

Fitz, J., & Halpin, D. (1994). Ministers and mandarins: Educational research in elite settings. In G. Walford (Ed.), *Researching the powerful in education*. London: UCL Press.

Flick, U. (2009). *An introduction to qualitative research* (4th ed.). London: Sage.

Gewirtz, S., & Ozga, J. (1994). Interviewing the education policy elite. In G. Walford (Ed.), *Researching the powerful in education*. London: UCL Press.

Guba, E. G., & Lincoln, Y. S. (1994). Competing paradigms in qualitative research. In Denzin & Y. S. Lincoln, *Handbook of qualitative research*. London: Sage.

Hawkins, K. (2002). *Law as last resort: Prosecution decision-making in a regulatory agency*. Oxford: Oxford University Press.

Kogan, M. (1994). Researching the powerful in education and elsewhere. In G. Walford (Ed.), *Researching the powerful in education*. London: UCL Press.

Lincoln, Y. S., & Guba, E. G. (1985). *Naturalistic inquiry*. Beverley Hills, CA: Sage.

Neal, S. (1995). Researching powerful people from a feminist and anti-racist perspective: A note on gender, collusion and marginality. *British Educational Research Journal, 21*(4), 517–531. https://doi.org/10.1080/0141192950210406.

Nelken, D. (2010). *Comparative criminal justice: Making sense of difference.* London (UK), Thousand Oaks (CA), New Delhi (IND) and Singapore: SAGE Publications Ltd.

Pearce, F. (1976). *Crimes of the powerful: Marxism, crimes and deviance.* London: Pluto Press.

Silverman, D. (2011). *Interpreting qualitative data: A guide to the principles of qualitative research.* London: Sage.

Snape, D., & Spencer, L. (2003). The foundation of qualitative research. In J. Ritchie & J. Lewis (Eds.), *Qualitative research practice: A guide for social science students and researchers.* London: Sage.

Walford, G. (2011). Researching the powerful. *British Educational Research Association* [Online]. Available from: http://www.bera.ac.uk/resources/resource-list?page=4&tid=All. Accessed 25 Aug 2014.

Westmarland, L. (2011). *Researching crime and justice: Tales from the field.* New York: Routledge.

Whyte, D. (2000). Researching the powerful: Towards a political economy of method. In R. King & E. Wincup (Eds.), *Doing research in crime and justice.* Oxford: Oxford University Press.

5

Enforcement Resources

This chapter analyses and compares the relationship between the resources available and the enforcement policies used by occupational health and safety (OHS) institutions, and the political mechanisms to distribute these for enforcement activities. This chapter concludes that justice is expensive and that austerity policies leading to the erosion of funds needed for regulatory enforcement activities have caused the decriminalisation of OHS crime and, hence, injustice. This chapter also raises fundamental questions on whether enforcement institutions should be provided with enough resources to fulfil a constitutional obligation to ensure justice, such as the case in the Italian legal system, or whether enforcement institutions' mandates should not be guaranteed, such as the case in Britain. The chapter is divided into four main sections. This first section reviews and compares OHS enforcement institutions' funding in Britain and Italy since the early 2000s. It then reviews the literature on how funding affects enforcement policies. The second and third sections of the chapter reports and analyses the findings of the qualitative interviews conducted with British and Italian enforcement officers. The last section critically analyses the main differences between

© The Author(s) 2019
D. Canciani, *The Politics and Practice of Occupational Health and Safety Law Enforcement*, Critical Criminological Perspectives,
https://doi.org/10.1007/978-3-319-98509-1_5

the British and Italian policies governing the distribution of resources among OHS enforcement institutions.

The Health and Safety Executive (HSE), the British OHS enforcement institution, witnessed a significant decrease in net operating costs from 1999/2000 to 2016/17 (see Table 1). At 2017 real price purchasing power parity (PPP), the HSE net operating costs between 1999 and 2016 have decreased by 47% from £289.9 million to £133.29 million. Real term resources fell every year on average by 3.9% between 1999 and 2016, 0.36% on average between 1999 and 2009, and 9.92% on average between 2010 and 2016. This sharp budget reduction from 2010 occurred due to the decision of the 2010 Conservative government to cut the HSE budget by 35% by 2015 (*Hazards Magazine* 2010). The introduction of the Fee for Intervention (FFI) policy from October 2012 has partially contained these budget cuts, which increased HSE revenues from £3 million in 2013 to almost £15 million in 2017. However, since the 2015–2016 financial year, HM Revenue and Customs has allowed the HSE to retain a maximum of £11 million only from the fees collected yearly. Hence, on average and at 2017 PPP, the HSE budget when including the funds collected from the FFI policy and the £11 million cap imposed in 2015, decreased by 8.88% between 2010 and 2016. This means that between 2000 and 2016 the HSE budget at PPP, which includes the revenue collected with the FFI, has effectively been cut by 48.8%. This trend has accelerated since 2010, and the reduced resources distributed to LAs from central Government since 2010 has also decreased LAs' capability to enforce the regulations. Due to the funding mechanisms used for quangos in Britain (but specifically in England), the HSE and LA funding cuts since the early 2000s have, effectively, been a major cause of the further decriminalisation of OHS crimes in Britain.

Since the LAs also have the duty to regulate OHS, it is also important to analyse the total expenditure of the regulatory services for LAs. The data is only available for England and only between the financial years 2009/10 to 2016/17 (see Table 2). Between 2010 and 2017 the net current expenditures at 2017 PPP of LAs on OHS enforcement has decreased by almost 40%, from £53 million to £31.5 million. This means that the net expenditures of LAs' enforcement activities have

Table 1 HSE budget and FFI (thousands)

Financial year starting	Actual annual net operating costs	Fees for intervention collected (included in net operating costs)	Yearly budget adjusted at 2017 PPP of relative real price of a commodity	Fees for intervention adjusted at 2017 PPP of relative real price of a commodity (included in net operating costs)
2000	£176,208		£281,900	
2001	£190,075		£298,800	
2002	£202,696		£313,400	
2003	£201,633		£303,000	
2004	£197,265		£287,900	
2005	£214,850		£304,900	
2006	£238,157		£327,500	
2007	£232,732		£306,900	
2008	£215,121		£272,900	
2009	£220,819		£281,500	
2010	£230,014		£280,300	
2011	£202,909		£235,000	
2012	£174,912		£196,400	
2013	£159,218	£2836	£173,500	£3090
2014	£153,863	£8706	£163,800	£8706
2015	£137,787	£10,150	£145,200	£10,700
2016	£134,129	£14,706	£134,100	£15,230

Source HSC (2002, 2003, 2004, 2005, 2006, 2007, 2008); HSE (2009, 2010, 2011, 2012, 2013, 2014, 2015, 2016, 2017a, b)

Table 2 LAs' net current expenditures

Financial year starting	Actual yearly net current expenditures	Yearly net current expenditures at 2017 PPP of relative real price of a commodity
2009	£43,488	£53,000
2010	£43,578	£50,480
2011	£40,087	£45,000
2012	£39,474	£43,000
2013	£37,721	£40,150
2014	£35,067	£36,960
2015	£31,984	£33,130
2016	£31,858	£31,858

Source Ministry of Housing, Communities & Local Government (2011, 2012, 2013, 2014, 2015, 2016, 2017a, b)

Table 3 Costs of SPSAL enforcement 2001–2009

	SPSAL total costs	SPSAL cost per head of economic active population (14–65 of age)
2001	€618,713,000	€18.31
2002	NA	€15.66
2003	€579,814,000	€11.36
2004	€503,527,000	€12.5
2005	NA	€13.3
2006	€399,490,000	€12.1
2007	NA	€14.01
2008	NA	€15.7
2009	€613,501,000	€15.87

Source Ministero della Salute (2004, 2006, 2007, 2009, 2010)

fallen by 7% per year (Ministry of Housing, Communities & Local Government 2012, 2013, 2014, 2015, 2016, 2017a, b).

The data of Italian OHS enforcement institutions financial resources are scarce and difficult to compare longitudinally (see Table 3). These figures should be interpreted with caution because the Ministry of Health has been collecting figures since 2001, but the different method-ologies adopted to collect data across Italian regions do not make these

Table 4 INAIL business support incentives (ISI)

Years	Euro
2002	€66,106,483
2003	Incentives NA
2004	€39,293,130
2005	Incentives NA
2006	€39,293,130
2007	Incentives NA
2008	Incentives NA
2009	Incentives NA
2010	€60,000,000
2011	€205,000,000
2012	€155,000,000
2013	€307,000,000
2014	€288,015,802
2015	€304,849,953
2016	€299,641,000
2017	€241,290,809

Source INAIL (2003, 2007, 2014b, 2016, 2017)

fully comparable (Ministero della Salute 2004, 2006, 2009, 2010). In addition, these figures are obtained by the Ministry of Health and, therefore, exclude regional funds. The activities and funds available to the Labour Territorial Inspectorates (ITL) enforcement institutions are not available. What emerges from the available data is that the funds allocated to Prevention Services for Safety in Work Environments (SPSAL) by the Ministry of Health decreased between 2001 and 2006, and by 2009 returned to 2001 levels (in absolute but not real terms). However, the figures show that SPSAL cost per head of economic active population has decreased since 2001 levels. This is due to an increase in the size of the economic active population.

Since 2006 Italian governments have passed legislation designed to increase funds for OHS. However, most of these funds have been directed towards National Institute for Insurance against Working Accidents' (INAIL) incentives supporting businesses, particularly small, medium and micro-sized enterprises (see Table 4) (INAIL 2003, 2007, 2014b, 2016, 2017). The increase in deterrence-driven enforcement activities and funding to SPSAL, as well as increased compliance-driven, or support activities witnessed since the early 2000s has been followed by a constant fall in incident trends. Since 1994, the SPSAL level of deterrence-driven proactive enforcement policies has remained constant at 5% of firms operating in the territory. These trends also demonstrate that in Italy there has been an increased emphasis on the need to increase the support given to stakeholders, but there has also been an increased emphasis on deterrence-driven enforcement policies. Despite the strict due-process practices of the Italian criminal justice system, OHS incident prevention remains a priority, which is achieved by a combination of SPSAL deterrence-driven policies and INAIL compliance-driven ones. This approach is, indeed, more expensive than in Britain.

The resources given to regulatory agencies can have an impact on the enforcement policies used, and thus the enforcement outcomes and the level of criminalisation of crimes. Most of the literature proposing innovative compliance-driven OHS enforcement policies ignore this issue. While conducting this research study, a fundamental difference between the way OHS enforcement institutions are funded in Britain and Italy emerged. In Italy, the regulator activities are dictated by constitutional values, criminal code and strict legal procedural enforcement principles, which means that the enforcement of the regulation is viewed as a service that the Italian state must provide to its citizens. This is not the case in Britain, where there is no written constitution obliging the state to provide a minimum level of regulatory enforcement. In addition, the British regulatory procedure is designed to use a certain amount of discretion when choosing the enforcement policies to adopt, but also when trying to achieve the regulatory goals. This means that in Britain the enforcement policies depend on the budget given to the regulator, while in Italy the budget given to the regulator is determined by the regulatory and enforcement objectives that are decided by Parliament, which must be compliant with constitutional values. This issue became particularly important in Britain between 2010 and 2015 when the HSE budget was reduced by a further 30%, which raised questions with regard to the minimum budget the British regulator can be given to achieve its statutory duty.

The resources available to an agency will have an impact on the enforcement policies adopted (Bartrip and Fenn 1980a, b, 1983) and a reduction in resources will cause the enforcement agency to use the most economical policies (Jost 1997; Posner 1972). This is much more likely to happen in a jurisdiction where the enforcement agencies have more discretion to choose the enforcement policies to use, which is the case in Britain, but not in Italy. In fact, resources in Britain can be more sensitive to political views and priorities and can be affected by the specific ideologies of the Government in power (Posner 1972; Spitzer 2000; Shavell 1982, 1999; Soderberg 2007). In a situation where the enforcement agency has the power to decide which enforcement policies to use, institutions such as the HSE engage in forms of cost-benefit analysis (CBA) to make the most efficient use of the resources available. In this case, however, by 'most efficient' it is meant that the HSE has reduced its commitment to prosecute, not because this is less effective than other policies, but because it is more expensive.

This was also confirmed by most of British pressure groups, trade unions and other organisations that submitted evidence for the Fourth Report of the House of Commons Work and Pensions Select Committee (2004) on the operation of the Health and Safety Commission (HSC). The Centre for Corporate Accountability (CCA) and trade union Prospect submitted evidence showing that falling resources cause a decrease in proactive and reactive inspections. They argued that the shift of resources from deterrence-driven to compliance-driven policies happened because the latter are more economical. The NHS Confederation also submitted evidence that the HSE lacks enough resources to meet the regulatory objectives within the NHS. The same opinion was echoed by many more organisations, such as the EEF Manufacturers' Organisation, trade union Prospect, and also by the same HSE head of the Field Operations Directorate (FOD) (House of Commons Work and Pensions Select Committee 2004). This evidence was provided before the 48.8% budget cuts experienced by the HSE since.

The Environmental, Transport and Regional Affairs Committee also argued that between 2002 and 2004 OHS prosecutions in Britain fell as a consequence of falling resources (House of Commons Work and

Pensions Select Committee 2004). Also, the CCA provided evidence that the decision to start prosecutions depends on the level of resources available, which has always been denied by the regulator. However, the HSE also has the right to seek compensation for prosecution costs, which means that resources might not affect prosecution levels as much as was thought. Despite this, the enforcement institution can still decide to decrease the number of prosecutions in order to aim to achieve a 100% success rate and avoid wasting resources on unsuccessful court cases. That is because losing prosecution cases can be seen as a waste of resources with no results in terms of regulatory compliance. Therefore, when resources are limited it is in the interests of the enforcement institutions to increase the rate of successful prosecutions, or increase the threshold by which a case is selected for prosecution. This, however, can only happen if the enforcement institution is given the power to take these decisions, as is the case in Britain, but not in Italy. Yet a commitment to increasing work effectiveness is a method to cope with lower levels of resources. In fact, the *Environmental, Transport and Regional Affairs Committee* report also argued that the number of successful prosecutions is also directly related to the level of expertise that the enforcement institution is capable of obtaining (i.e. paying for) (House of Commons Work and Pensions Select Committee 2004).

Another piece of evidence presented by the *House of Commons Work and Pensions Select Committee* (2004) report is that falling resources were causing the HSE to devolve more of the limited resources available to more economic activities, such as compliance-driven enforcement techniques. Trade union organisations, such as the Trades Union Congress (TUC) and GMB, the general trade union in Britain, reported that the HSE used resources dedicated to proactive and reactive enforcement activities to offer free-of-charge printed information for duty holders. Although offering free-of-charge printing materials for duty holders is important, the resources needed to undertake these activities should not be transferred from other deterrent enforcement activities, such as proactive inspections and prosecutions. The same argument was made by Royal Society for the Prevention of Accidents (RoSPA) with regard to the prevention and enforcement of work-related OHS road incidents. These are investigated as OHS-related incidents in Italy, but not in Britain (House of Commons Work and Pensions Select Committee 2004; INAIL 2014a).

In previous empirical research where enforcement officers were interviewed, resources were never mentioned as an issue affecting the work of the OHS enforcement officers (Hawkins 1984, 2002). Ayres and Braithwaite (1992) did not take into consideration the issue of resources when analysing OHS regulatory agencies. In Baldwin and Black's (2008) really responsive regulation enforcement theory they claim that enforcement policies should take into consideration a wider number of contextual characteristics, but they fail to take into consideration the critical issue of funding. Only Tombs and Whyte (2010) and Davis (2004) speak extensively about the fall in real-term resources that occurred to the HSE in the 2000s, and the various domino effects this erosion has caused to the activities of the HSE.

The level of resources available to the judiciary can also have a direct effect on enforcement levels and court sentences, especially in Italy where the Judiciary Authority bears most of the prosecution costs. The funding available to the Italian Judiciary Authority, however, is low. Tinti (2007) argues that Judiciary Authority's funding has been inadequate for a long time and especially for those defendants that are represented by well-paid and well-prepared lawyers. Indeed, as an Italian journalist argues

> There is no rhetoric […] in the daily lives of magistrates [public prosecutors], who, together with police officers, Carabinieri, financial police [and] administrators, every day attempt to manage the Judiciary systems in spite of the saints, namely parliamentarians, Governments and maybe also a good section of the population, who are only terrified of a functioning justice system. (Travaglio 2007, p. XII)

This indeed complies with critical criminologists' viewpoints, who see the criminal justice system as a means to protect the interests of the ruling classes (Pearce 1976; Jones 2013; Reiman and Leighton 2010).

A weakness in the literature is that there are few testimonies of the issues faced by enforcement officers when faced with low resources, but also a lack of the different views of officers in different jurisdictions. If the discretion to choose enforcement policies allows the enforcement institutions to substantially change the nature of their work, it is interesting to know what happens in jurisdictions where the enforcement institution does not have this freedom, such as in Italy.

Table 5 Interviewees who mentioned lack of resources as an issue

Country	Total interviews	Interviewees mentioning lack of resources	Percentage of interviewees mentioning lack of resources (%)	Total times interviewees mentioning lack of resources
Britain	9	7	77.7	11
Italy	13	11	84.6	30

Out of a total of 22 interviews, 18 enforcement officers talked about resources (see Table 5). During interviews, enforcement officers mostly talked about financial and human resources. However, interviewees also talked about resources in general terms. That is because resources, whether financial, human or equipment, usually decrease and increase together, and so interviewees did not differentiate between them.

Three main themes emerged from the interviews. The first theme explores the significant number of complaints about the scarce level of resources available. The second theme concerns the evidence supporting enforcement officers' complaints about the scarcity of resources.[1] In order to understand whether lack of resources is effectively a hindrance for enforcement officers, interviewees were asked to justify their arguments. The third theme explored relates to the effects that scarcity of resources has on enforcement activities. The following sections explore, respectively, the British and Italian OHS opinions on these themes. The last section summarises and compares the opinions of the OHS enforcement officers in Britain and Italy.

1   Britain

Seven out of nine enforcement officers interviewed in Britain complained about resources or argued that more resources would have helped in their enforcement activities. A number of enforcement officers made specific links between the human resources available and their

---

[1]It is useful to note here that resources can be a relative concept, i.e. they are not infinite and enforcement institutions can always put more resources to good use.

ability to accomplish their work successfully. While talking about the balance between the work activities conducted in the office and those in the field, a British enforcement officer argued that

> With the current staffing level the amount of advice that we are called to give has gone down because the number of inspectors has dropped. Currently, I would say, in the last year or so I spent probably about 70% of my time in the office. Occasionally most of my work has rarely been to do with investigation work, which means a lot of office work, a lot of following up, and lots of procedures in terms of evidence collection and so on. (British interview 1, Line 33)

Later on, after mentioning the drop in the number of enforcement officers that occurred since 2000, the same enforcement officer also explained the following.

> Yes, I think it's one third down. It is quite a lot since I joined the Health and Safety Executive, which doesn't necessarily help really. […] What gets measured is actually what is countable in these days, on computer systems with Government targets. It is the target that is actually the most noticeable things about this [New Labour] Government, everyone is looking at targets and statistics and we are maybe losing the big picture really. It is quite frustrating really… Because we are down on the number of enforcement notices which I think it's probably largely due to the number of drain of inspectors. Less inspectors and we are maybe less able to go out as much…. (British interview 1, Line 673)

This interviewee linked the fall of enforcement officers that occurred since 2000 to a decrease in enforcement activities, which might range from improvement notices to prosecutions. These two statements confirm that enforcement activities take up a lot of time. The question is whether fewer resources results in less inspections.

During the interview officers were asked whether the lack of resources leads the enforcement officers or the institution employing them to choose enforcement policies and practices that are inexpensive. This argument emerged three times from the interviews. In fact, officers were asked whether deterrence-driven enforcement policies are less likely to be used because they are more expensive. The first British enforcement officer to talk about this said:

> [They] are getting too pally with industry and we are using other tools, rather than saying, if staff numbers have dropped by 20% then, hey, it is not surprising that formal enforcement is dropped by about 20%. Each inspector is there. [...] in fact I think that inspectors are using formal enforcement more now than they have in the past. Each inspector is actually serving more notices now than they have ever had in the past. (British interview 7, Line 562)

Leniency is a consequence of fewer enforcement activities and officers, but also caused by issuing fewer improvement and prohibition notices.[2] That is because they do not have as much time to go back to firms as often as they used to in their advisory roles, which means that issuing notices makes duty holders' compliance quicker, and so more financially efficient. Taking prosecutions is more time consuming than issuing improvement and prohibition notices, which means that less resources has also meant fewer court cases. This finding demonstrates that greater use of deterrence-driven policies is more efficient and effective, but also more expensive. Prosecutions are not used as much because they are too time-consuming. Another British enforcement officer made this point clear when mentioning that

> You would serve a notice because certain [hazardous] conditions have been met. You would prosecute because certain other [hazardous] conditions have been met. You would not prosecute because you found, as said earlier, dangerous machinery. You would serve a notice, that is what

---

[2]It is important to note that informal enforcement involves those actions concluding in informal or personal advice, while formal enforcement is when officers issue improvement or prohibition notices and prosecution. However, at the time of the interviews in England notices did not require firms to pay penalty fines. The only time when firms might have been penalised, either with fines or through employer imprisonment, is when firms were taken to court for prosecution. This has now changed with Fees For Intervention. Improvement and prohibition notices represent the formalisation of informal requests. Hence, until the Fees For Intervention was introduced the primary enforcement threat of prosecution to firms was when they were refusing to comply with improvement or prohibition notices, which a very rare occurrence.

the notice is there for. I don't think that it has got anything to do with resources. [...] with declining resources, I think that you are more likely to serve notices, because you don't have the time to keep going back in an advisory capacity. So, resources would have a direct effect on that [serving notices]. (British interview 6, Line 307)

While asking another enforcement officer whether enforcement policies that placed a stronger emphasis on punishment would change the practices of the enforcement institution, the interviewee responded as follows.

> *Interviewee*: I think that we would be doing, we would be serving more notices, we would be doing more prosecutions, but the more of those that we do, the less time we actually have to go out and doing inspections because they can absorb an awful lot of time particularly in large prosecutions. So, it is actually in my view better to start off by providing advice and then escalate up if you need to because providing advice during inspections, if you are physically there, if you have got the leaflets [information] with you it is easier than having to take a prosecution, [for actually might be] not a particularly serious breach of the law, but that takes an awful lot more time.
>
> *Interviewer*: So, you are saying that the advising role might also be considered to be less burdensome?
>
> *Interviewee*: Yes... I think it is, but also is potentially quicker and more cost effective because it takes less time to advise someone to give them a leaflet and having a conversation with them than it does to prepare all of the paperwork for a notice or to go to court. And so, within a day I can advise more businesses than I can prosecute. So, I can see more people and I can have more of an impact because I can protect as many employees as there are in ten businesses by advice than I can deal with by prosecuting one business. And if I am prosecuting someone I am not protecting employees. What I am doing is punishing the business for what they have not done in the past. (British interview 9, Line 45)

This enforcement officer recognised the policies as functional to the need for improving safety, but also justified the reasons to adopt them on the basis of a lack of financial resources. This is evidence that resources affect the enforcement activities used by officers and the enforcement institution. Three significant messages emerged from the quotes above. Firstly, interviewees argued that enforcement practices, such as issuing notices and starting prosecutions, are activities dictated by specific enforcement procedures, but that they can choose whether to issue notices and reduce or increase the time spent on each firm. Hence, enforcement officers have a certain degree of freedom to decide whether just to adopt an advisory role or issue an improvement or prohibition notice and later return to site to check if the firm has complied with it.[3] This demonstrates that although the enforcement officers are there to execute rather than formulate policies, their rationales for choosing which enforcement policies to use might also be influenced by the level of resources available (such as time in this case) rather than only being based on the Enforcement Management Model (EMM) and Enforcement Policy Statement principles.

Secondly, enforcement officers have the flexibility to decide whether to formalise their enforcement actions, but only to the point where they think that a breach deserves prosecution. The reason for this is that the more formal the enforcement action, the more complex and risky it becomes and the more resources it might require. Therefore, enforcement officers and the institution engage in forms of CBAs to assess on a case by case basis whether it is worth escalating to more punitive enforcement actions. The decisions are complying as much as possible to the EMM, Enforcement Statement principles, and other protocols. However, resources are almost as important when taking these decisions, which is understandable. The enforcement institution also faces the challenge of predicting the resources needed to conduct enforcement activities until the end of the year, while trying to make their enforcement activities as consistent as possible. These decisions are taken mostly at institutional level, but officers also reflect on these issues in their day-to-day activities.

---

[3]In a different context, one of the respondents argued that some counties register no notices in a whole year, which is very difficult to happen because, according to another respondent, in 99% of workplaces there are improvements to be made.

In the final quote, the officer explained that the reason why prosecutions are not as effective in ensuring compliance as advice and enforcement notices is because they are very lengthy and expensive processes. The enforcement officer does not argue that prosecutions are less effective tools for ensuring compliance, but that advice and notices, and collecting fees for interventions, are not resource-intensive and are therefore used more often than prosecution. This is evidence that fewer resources affect enforcement activities and outcomes negatively. This might contribute to the under-criminalisation of OHS crimes because budget changes, together with enforcement officers' discretion, have a deteriorating effect on the number of deterrence-driven enforcement activities conducted.

In another interview, an enforcement officer argued that resources were not affecting their work activities, but when asked about the changes that the 30% reduction in the HSE budget might have on the enforcement activities after 2015, the interviewee argued that "it will change the nature of what we do" (British interview 2, Line 355). A few minutes later the same interviewee explained that by '*the nature*' they meant new opportunities, and new policies, such as the new FFI policy, which was announced and introduced several months later. In an earlier instance the same interviewee, while arguing about policies, explained that the enforcement institution needs to use persuasive techniques with industry associations because "there is no way that we are going to get inspectors to visit every single one of those employers [...]. We need their industry association to make that leap forward" (British interview 2, Line 247). In this specific case, the officer implied that enforcement activities conducted depend on the human resources available. Widespread and well-settled firms are easier to reach compared to small, medium and micro enterprises, but this means that without help from industry or a decent level of resources it is impossible to reach out to small, medium and micro enterprises. Hence, cooperating with trade and industry associations is essential to achieve OHS compliance in these firms, especially when enforcement resources are scarce. Enforcement policies that try to *convince* trade and industrial associations to encourage compliance with the OHS regulations is important, but these policies should be supported by a tangible enforcement actions from the enforcement institutions. The current level of human resources makes it impossible to provide sufficient enforcement activities to achieve these goals, and cooperating with industries is a good and economical alternative, which OHS enforcement institutions cannot refuse to consider.

While talking about prosecution processes, a number of enforcement officers complained about the low numbers of OHS-related prosecutions and penalties. A British enforcement officer argued that

> [They did] not think that the cost of taking prosecution has directly been an issue to determine whether we do it or not. The issue is the legal processes. The legal procedures to actually get a case into courts have become much more complicated over the course of my career. So, it is now much more time-consuming to get a case to court. Nobody has ever said that 'you must not take that case to court because we cannot afford it', but the reality is that we only have so many staff, we only have got so many working hours and it is only so much time they have actually got to do it. So, I suppose, logically, one would have said: we will increase our total number of staff proportionately to account and relate to the complexity of the process and get the cases to court. That has not happened, but no one has ever said that we cannot afford to do it. (British interview 3, Line 1123)

It is interesting how on this occasion the enforcement officer argued that while the enforcement institution has not actively stopped prosecution cases, fewer resources have also caused fewer prosecution cases to happen. It must be considered that an enforcement officer's ability to carry on with a prosecution case also depends on the enforcement institution's ability to provide resources and assistance. The enforcement officer seems to be aware that the lack of resources (mainly human) can cause a reduction of prosecutions. This might mean that the interviewee is aware that OHS in firms is under-enforced. However, shortly after this statement the same enforcement officer explained

> In the comment I have just made, I was probably thinking back twenty years. I mean, there are procedures in place now that if any one group of inspectors, any operational group is totally overloaded they have the incidents and accidents that according to the criteria laid down should

be investigated and they have not got the resources to do it, then, that information will be passed up to the more senior manager who would normally be expected to find resources from somewhere else within his command, divert resources to do it. So, we maintain this reality of fairness and consistency in our approach. I mean, there are inspectors who do get swamped almost full time. If you get a high profile difficult prosecution or more than one, people can be occupied for a year or two doing almost nothing else other than that work. (British interview 3, Line 1147)

The officer argued that the enforcement institution considers each potential prosecution on a case-by-case basis, and if needed finds resources to ensure that necessary enforcement actions are taken. This is to ensure that principles of fairness and consistency are met. This demonstrates that there is a direct correlation between resources and enforcement activities, and the HSE needs to strike a balance between the resources available and the aim of achieving the organisation's basic enforcement goals. This might be considered a serious institutional weakness of the British OHS regulatory enforcement institution because the enforcement institution freedom to choose the enforcement activates according to the resources available only ensures the basic principle of equality before the law partially, and this can cause the under-criminalisation of these crimes. This system implies that ensuring a rational use of resources is more important than enforcing the regulation and ensuring an adequate level of OHS, or justice, to citizens. Budget decisions are political and ideological decisions which set priorities, but if the priority is to allow firms to kill and injure workers without penalising them for it, the system of justice highlights an issue of social inequality that is embedded into the British legal system.

On a different occasion, while arguing about the balance between making a cost-efficient use of resources versus enforcing the regulation effectively, a British enforcement officer argued that

You can always use more resources and you can always put more resources to good use because it is not a bottomless pit. I think that a few years ago the resources were adequate. We needed to concentrate on efficiency and using the resources that we had adequately. I suspect that resources are generally tight now, because a lot of resources have been stripped out.

[…] So, I guess that the overhead is slightly higher than the end of an efficiency drive per inspections that it was at the beginning. Whereas, if you could have more inspectors the overheads would not go up proportionally. So, it is not easy to survive [the Government's] cuts without keeping the efficiency as high as it was. I think that it loses things, however hard you are trying to gain things. So, I suspect that at the moment resources, perhaps, are more of an issue than… I mean everybody always say that they need more, because they could always use more and you could put them in good use, but I think that probably is the case now that resources are limiting what can be done. (British interview 4, Line 545)

According to this interviewee the 2010 government's decision to reduce HSE funding by 30% by 2015 might have been a step too far for OHS enforcement institutions. In addition, the revenue collected through the FFI policy between 2012 and 2017 has only compensated for 30% of the budget cuts introduced since 2011. Including the funds collected through the FFI policy, since 2000 the HSE, at PPP, lost 48.8% of annual resources. An enforcement officer who was keen on preserving his anonymity argued that decisions about the enforcement institution funding are political, and that it is not surprising that the number of OHS enforcement officers on stress-related leave increased in 2011.

2   Italy

Italian officers also complained about the lack of enforcement resources available. They argued that the human resources available are scarce and that this has an adverse effect on their enforcement activities. One of the arguments that emerged from the interviews is the balance between the national enforcement targets and resources. An Italian enforcement officer said that

The institution of an enforcement and prevention system in work environments that, more or less, can reach 5% of the national firms per year means that I have the probability to visit one firm roughly every twenty years. It is very well-known that in certain sectors… In construction I can set up a site, commit health and safety breaches, close it, and disappear because the probability of receiving a visit from us is extremely low. If we visit these types of firms once every ten years the penalty that we ask the employers to pay is nothing compared to the amount of money that the firms have made in the past ten years. This system is unacceptable because it gives a competitive advantage to the firms that do not comply with the regulations. [...] So, it would be very useful if the enforcement officers could have a little bit more resources for enforcement. But this is a choice of the state, the state says where the resources go, but I do not want to talk about this. I can only say that there should be a wider diffusion of these structures and more resources because we are monstrously too few in comparison to the needs. (Italian interview 11, Line 294)

The interviewee complained about the lack of proactive enforcement activities, which is caused by a lack of resources. A lack of financial resources decreases the chances of detecting non-complying firms, especially transitionary or temporary micro enterprises. It is very interesting that this officer also argues that the lack of resources can be linked to the role of the (Italian) state political interests. In other words, the 20-year average inspection frequency per firm, and the under-criminalisation this causes is attributable to a political institution which is principally driven by capitalist interests. Interestingly, only on one occasion a British enforcement officer, who did not want to be recorded, argued that these are political decisions driven by the same interests. A similar opinion was also given by a colleague when asked whether the national enforcement target of 5% is sufficient.

Technically it is impossible to do more. It [5%] is not sufficient, but is impossible [...] because it is a very high number. Because 5% represents a number of firms that at the moment we can just about visit. (Italian interview 9, Line 184)

This officer argues there is a willingness to see an increase of enforcement activities, but only if the SPSALs receive more funding. On a different occasion, while talking about the preparation and on-the-job development of enforcement staff, an Italian enforcement officer argued that

We do not have the funds, we do not have the money to develop our officers. There are twenty-two officers and we have twenty-two hundred euros to train them every year. No way close to being enough. […] We need money to train our officers, because when we enter into a firm our officers know ten and the firm's consultants know twenty, if not hundreds. So, there is an enormous gap in the technical knowledge of our officers. (Italian interview 11, Line 550)

This is evidence that resources are too limited to improve officers' technical knowledge and be more efficient in their daily enforcement activities. This issue creates major difficulties when dealing with private organisations, where professionals consultants might be better prepared than enforcement officers. While talking about the same issue, another Italian enforcement officer argued that

The principal and significant obstacle is surely the number of operators, hence the human resources available. Too few. Every year… We get to the end of the year, we reach the annual targets, but our tongues are on the floor.[4] Now to do a good inspection you need time. To do a good inspection you need to be prepared, hence updated. To do a good inspection you need to have the serenity to tackle well and with enough time different aspects of many issues. Clearly by starting to deal with numbers while being short of staff we risk being superficial, and hence obtaining a low achievement in terms of health and safety prevention. (Italian interview 5, Line 177)

This is another example of an officer linking the resources available with the quality of the enforcement work done, and how low resources can negatively affect the practical implementation of enforcement policies and activities.

One of the aspects that the Italian enforcement institution relies on is also the employment of external consultants to deal with specific situations. While talking about the need to access high levels of professional expertise in specific situations, an Italian enforcement officer argued that

There is the ISPESL, but I have tried several times to call the engineers, because they are supposed to help you in practical cases, but when you

---

[4]Italian expression meaning 'to be exhausted'.

phone them there is no reply.[5] So what are you supposed to do? You go through informal channels. Someone you know, or you approach the magistrates [public prosecutors] and tell them 'you should call an expert'. Then the magistrate asks me if I know an expert on the subject. You give him a name and [the magistrate] nominates them, but this also happens less often because the Judiciary has always got less money available. (Italian interview 1, Line 892)

A lack of expertise and resources prevent officers from attaining information that can lead to a successful completion of incident investigations and prosecutions. The issue is that the Italian Judiciary Authority also lacks resources. Hence, the statement depicts a situation where both the enforcement institution and the Judiciary Authority try to combine resources in order to prosecute OHS breaches successfully.

This issue was also highlighted in another interview when talking about the Thyssen foundry tragedy in Turin, where six workers died in an explosion in December 2006. OHS enforcement officers had visited the site shortly before the incident. An Italian enforcement officer argued that

Those poor inspectors that went to the Thyssen [before the explosion] were not dishonest people; they were people that did not have the instruments to prevent the accident. They did not have them, they would have never had the possibility, the capacity, and the technical capability to enter and assess the plant. (Italian interview 11, Line 339)

The issue has also been highlighted in practical terms during the court case as the worst Italian OHS tragedy in recent times. Hence, the interviewee suggests that with more resources, personnel and instruments the chances of the incident happening might have at least decreased.

---

[5]ISPESL, Superior Institute for Prevention and Work Safety, was founded as the Italian research centre for health and safety hazards and its practices were meant to offer information, expertise and support to health and safety inspectors. ISPESL was shut down in 2010 and responsibilities transferred to INAIL. The amount of help that ISPESL was offering to enforcement officers was criticised also in other interviews.

Another issue that emerged from the interviews was a strong willingness among interviewees to increase proactive enforcement activities, but the inability to do so because of a lack of resources. An Italian enforcement officer, while talking about the reasons hindering enforcement activities, argued that

> The limits obviously are those of the availability of an adequate number of personnel and resources, in short, to do it [proactive enforcement]. With effort we can guarantee our presence at 5% of firms with employees. [...] If we would translate this 5% of the total number of firms, it means that a firm will not be visited for at least ten to twenty years. (Italian interview 4, Line 89)

This is another example of an officer who would like an increased the number of enforcement officers on the territory, which is dependent on the level of resources available. Later on, the same interviewee explained the complexities of how the resources available are linked with other vital aspects of OHS enforcement.

> If we have an adequate level of resources, we can definitely increase our presence in the territory. The limits are a little bit linked to some legal rigidity, which nowadays are oriented towards firms that can afford to have an internal system of [OHS] prevention. While the majority of the economic reality on our territory, but also in the whole of Italy, is constituted by small or micro firms where the architecture of Legislative Decree 81 2008, but also of Legislative Decree 626 1994, is implemented with a lot of difficulties. (Italian interview 4, Line 100).[6]

Thus, a lack of resources prevents officers from increasing the amount of information given to micro, small and medium-sized enterprises. Although goal-setting legislation is very appreciated among OHS enforcement officers in both countries, this interviewee also argues that duty holders—and particularly SMEs lacking resources—are struggling

---

[6]Legal Decree 626 1994 (Repubblica Italiana 1994), which in 2008 was replaced by Legal Decree 81 2008, was the first regulation to introduce the goal setting philosophy in the health and safety Italian law in the same way the Health and Safety at Work Act 1974 did in England.

to understand and implement this unique and innovative legal framework. Hence, this quote is evidence of one of the main difficulties encountered by OHS enforcement officers as a result of the introduction of goal-setting legislation. Namely, the necessity to assist micro, small and medium-sized enterprises to set up and maintain internal systems of OHS management based on risk assessments.

Another issue linked to resources regards the availability of technical resources, such as instruments required for chemical analyses during enforcement. An Italian enforcement officer argued that

> Also [technical] instruments are important. Obviously we have them, but they should be updated constantly because the technologies progress. An instrument of five, six, seven years ago is not good anymore for the regulation. Hence, resources to be used for technical instruments. From Legislative Decree 626 1994 our enforcement activity has changed radically. We must now also assess the risk assessment that the employer has done for the work environment. Hence, to assess the accuracy of the risks assessments, we should also be able to measure determined physical risk factors, such as noise, vibration, micro-climates, electromagnetic fields and so on, to see how well the employer has done the risks assessment. To do this we need instruments, but these are not enough at the moment. Hence, human and technical resources are fundamental to prevent and to suppress [OHS breaches]. (Italian interview 12, Line 156)

This quotation highlights a connection between the enforcement changes that occurred with the introduction of a goal-setting regulation and the need for technical instruments to conduct inspections and investigations. The last two statements indicate that a goal-setting regulation together with lack of resources favours large firms, hinders small and medium-sized enterprises from complying, and drives the enforcement officers to prefer compliance-driven over deterrence-driven policies. It is noteworthy to mention how these policies have changed the state's welfare recipients from citizens, through the provision of social security and benefits, to enterprises, through the provision of help to firms. Hence, advice and business consulting services offered by the enforcement institution are policies causing deregulation of the economy, which occurs by decreasing the rights of workers and increasing those of the employers.

During the interviews, it emerged that the Italian officers also engage in forms of CBAs while thinking about enforcement actions, which is induced by the need to maximise the use of resources. In fact, an Italian enforcement officer explained that

> Resources will never be infinite, but we have to understand once and for all how resources should be used and how much resources are really needed to support the system. Resources have to serve quantity and quality goals. So, increasing the quantity without changing the quality is a colossal mistake. So, we have to question the effectiveness of the interventions. Because if a hundred interventions result in a hundred useful outcomes we know that the quantity of resources deployed cannot improve [because their effect is financially efficient]. But if in a hundred interventions only 50% result in a useful outcome it means that before trying to increase the level of resources available I have to work on improving how the resources are used. (Italian interview 7, Line 116)

Once more, the level of resources available is paramount for conducting enforcement activities. The interviewee also engages in CBA calculations of enforcement policies and the results they can achieve. It is interesting to note that, when asked with closed-ended questions, interviewees denied using CBAs to decide enforcement policies. It is difficult to understand the extent and frequency to which officers undertake these calculations, but it happens, and this might influence their decisions while planning and conducting enforcement activities. This quote also illustrates another issue that was also mentioned by a British officer, namely the difficulty in establishing a causal qualitative link between enforcement practices and the OHS improvements achieved in the workplace.

3   Summary

This chapter has analysed enforcement officers' opinions on the resources available to enforce OHS regulations in Britain and Italy. The literature that has influenced the British enforcement policies since the end of the 1990s does not analyse this issue in depth (Hawkins 1984, 2002; Ayres and Braithwaite 1992; Baldwin and Black 2008). Only Tombs and Whyte (2010) link the consistent fall in resources that occurred in Britain and its effect on enforcement policies. In Italy, no one has ever analysed the statistics available and the opinions of the enforcement officers on resources.

Financial resources have fallen in Britain by 29% between 1999 and 2011 and by 35% between 2011 and 2015 (Tombs and Whyte 2010). By including the revenue collected through FFI policy, the total reduction of the HSE budget between 1999 and 2016 decreased by 48.8%. Italian statistics suggest that while resources decreased between 2001 and 2006, by 2009 these returned to 2001 levels (Ministero della Salute 2004, 2006, 2007, 2009, 2010). In addition, from the enforcement data available, OHS enforcement resources in Italy might have increased at least until 2011. Therefore, by summer 2009 (when the interviews were conducted), Italian OHS enforcement funding had been increasing for two years consecutively.[7]

From the interviews it emerged that there is a consensus among British and Italian OHS enforcement officers; low levels of funding and human resources affect their operations and put their enforcement activities under pressure. Italian interviewees complained about resources more openly and frequently (three times as much) than their British colleagues despite the increase in available resources over the years preceding the interviews. The reason for the higher number of complaints might be due to the freedom of Italian civil servants to criticise government policies, a freedom that their British colleagues do not seem to enjoy to the same level, or due to the specific experiences of the enforcement officers in the regions where the interviews were conducted. Given that the Italian OHS enforcement institutions have less freedom to decide how to spend resources when compared to their British colleagues, it also means that they have less flexibility to adopt more economical enforcement policies. It is the Italian Parliament that decides the level of funding. During economic crises, such as Italy has been experiencing in particular since 2008, it is also in the Parliament's

---

[7]According to a Ministry of Health report, in the regions where the research study was conducted funding increased by 33% between 2001 and 2009, most of it (17.8%) between 2007 and 2008 (Ministero della Salute 2004, 2006, 2007, 2009, 2010).

interest to try to reduce expenditures. These interests are, however, contraposed by the electorate's demand for greater protection and the constitutional requirement that guarantees a minimum level of enforcement activities. In other words, while in Britain an enforcement institution's goal is to achieve as much enforcement activity with the resources available; in Italy the goal is to provide enough resources to achieve the enforcement activities dictated by the electorate's demands and constitutional values.

Hence, several final conclusions can be made. The resources to conduct what British and Italian officers consider appropriate OHS enforcement activities are scarce. All interviewees expressed a level of distress about their workload and the pressure to perform adequately with the resources available. This was the case for both the British and Italian enforcement officers. Throughout the interviews, most interviewees complained that lack of resources resulted in ineffective enforcement activities. However, most of the British and Italian enforcement officers did not see resources as being a political issue. Only one Italian and one British officer argued that lack of funding is a political decision, in the sense that they perceived their OHS regulatory enforcement role as antagonistic to economic and political national interests. A British interviewee argued that the number of OHS enforcement officers on stress-related leave increased in 2011, which was the year after the HSE started to reduce its budget by 30%. The situation in Britain worsened so much that even officers, who are supposed to enforce stress-related breaches, have fallen victim to the same issue. Therefore, interviewees confirmed the concern of Tombs and Whyte (2010) that resources are falling and that they are inadequate for undertaking enforcement activities effectively.

The responses also show that both British and Italian enforcement officers engage in CBAs when analysing the enforcement policies they use. There are two main differences between the two jurisdictions. In Britain, the discretion embedded in enforcement policies used means that enforcement officers and institutions can run forms of CBAs and might adapt the enforcement policies to suit the level of resources available. In Italy, the prohibition on using discretionary practices, and the strict legal framework in which inspectors must operate, means that it is more difficult for enforcement officers to adapt practices to the resources available. Therefore, lack of resources together with enforcement officers' discretionary practices, as used in Britain, can lead to more financially efficient enforcement activities, but also to their decline and under-criminalisation of OHS crime. Lack of resources together with enforcement officers' and enforcement institution's non-discretionary practices, as used in Italy, means that resources cannot be reduced excessively because the state must guarantee a minimum and constant level of OHS enforcement activities. This, however, increases the costs of OHS enforcement. Arguably, the funding system in Italy distributes a more constant level of resources to the OHS enforcement institutions than in Britain. That is because the operations of the Italian enforcement institutions are decided in Parliament and guaranteed by constitutional values. This system is designed to ensure more pluralist decision-making and more constant protection of workers' rights, regardless of the state of the national economy or other international economic forces and agreements.

Deterrence is created by the likelihood of detection and the level of penalties issued (Becker 1968). The likelihood of detection can be significantly affected by the level of funding available to the enforcement officers. The level and frequency of penalties, or enforcement charges, might help to recover some of the costs of enforcement, but in both countries these have not been issued as frequently as expected. In Italy, penalties have been adopted as an enforcement tool since 1994, while in Britain these have only been introduced since 2012. The level of penalties issued after court prosecutions has increased in Britain in 2013 after the new sentencing guidelines was introduced, although its practical effects are yet to be assessed. From the interviews it emerged that sanctioning OHS crimes with high penalties was considered counter-productive for the achievement of regulatory aims. The reason for this is the so-called deterrence trap (Coffee 1981; Fisse 1983), which is encountered when sentences are designed to avoid bankrupting firms rather than employ the concept of 'just deserts'. Bankrupted firms cause job losses and employees become victims of the regulation that is meant to protect them. Hence, defendants can decrease penalties by threatening the prosecutor and the court that they will declare bankruptcy if penalties are too high. This means that even with the new sentencing guideline, firms might rarely be penalised adequately. These issues might be much more felt during economic downturns or when unemployment rates are high. Consequently, officers consider prosecutions and higher penalties as an ineffective and inefficient means to enforce the regulation. Hence, to obviate this issue and increase deterrence, proactive enforcement activities should increase in frequency. Therefore, more financial and human resources paired with low penalties are essential to increase deterrence and achieve compliance. Cutting resources can have a significant impact on the deterrence of the law and the efficiency of regulatory institutions' activities.

On-the-spot penalties were considered an essential enforcement tool to increase deterrence, penalise non-compliers and encourage a regulatory compliance culture, but only if the threat of detection is high. Enforcement of on-the-spot penalties, or enforcement charges, thus, have key roles in regulatory enforcement activities. According to Becker's deterrence theory (1968), increasing the number of enforcement officers without allowing regulators to issue on-the-spot penalties or charges makes inspections a very ineffective and inefficient deterrent enforcement tool. This means that until 2012, before the FFI policy was introduced in Britain, even if the number of British enforcement officers increased, their deterrent effect would have not improved significantly. In Italy, breaches have been penalised by compulsory administrative on-the-spot penalties since 1994 and the generally stable level of resources, when compared to Britain, for field operation means that the enforcement institution has managed to increase its deterrence power at little or no extra cost. Enforcement notices are a quick and financially efficient means to ensure compliance because employers have an incentive to comply with the regulation *before* inspections, and officers can prevent the occurrence of incidents with minimum expense and an immediate return in terms of safety. While the FFI policy increases the efficiency and effectiveness of British officers' field operations, the estimated 48.8% adjusted decrease in HSE funding between 1999 and 2016 means that the enforcement institution has witnessed a reduction of enforcement activities, and hence of the deterrence power they could exercise. By adopting on-the-spot penalties, an increase of enforcement officers might have a direct positive impact on compliance levels. Therefore, until 2012 an increase in number of enforcement officers might have only caused a trivial increase of deterrence on firms, which means that funding cuts to regulatory enforcement institutions have a significant negative impact on the level of deterrence and the effectiveness of their enforcement operations.

Resources also affect the number of prosecutions conducted, as well as the rate of guilty sentences reached. The main difference between Britain and Italy is that while in Britain lack of resources determines the under-criminalisation of OHS crime at enforcement institutions stage, in Italy lack of resources causes their under-criminalisation at prosecution stage. The British and Italian prosecution systems work differently to each other. In the former the OHS enforcement institutions are not required, with a few exceptions, to refer cases to the Crown Prosecution Service (CPS) and have the discretion to decide whether to take cases to court according to various enforcement agreement and procedures. In the latter, penalties and prosecutions are compulsory and only public prosecutors can decide to discontinue a prosecution case and only if the criminal evidence available is judged insufficient. Hence, prosecutions in Britain are considered an effective means to ensure compliance, and pursued only if enforcement institution has high chances of winning in court. There are various guidelines used to take these decisions, such as EMM and the Enforcement Policy Statement and protocols agreed with the CPS, but given the discretionary power of the regulator, it seems inevitable that the resources available also have a high chance of affecting these decisions. In Italy, this discretion is not available, but the success rate of prosecutions (estimated by an interviewee to be at around 50%) is much lower than in Britain (above 90%). Hence, regulatory institutions' resources must only have a minimal impact on the decisions taken by the enforcement institutions and the Public Ministry (PM). In Italy, the successful rate of prosecutions depends on the quality by which evidence is collected, but also on the expertise and knowledge that public prosecutors and judges have on the technicalities of the OHS regulations and requirements, which was often judged inadequate by officers. Therefore, the Italian interviewees argued that dedicating more resources to train enforcement officers with better legal skills and training public prosecutors, lawyers and judges to understand specific OHS-related matters should improve prosecution rates and decrease the under-criminalisation of these crimes. However, this is unlikely to happen given the costly and overburdened condition of the Italian legal system (Nelken 2010; Tinti 2007).

Thus, it can be argued that justice is expensive in both countries, but only victims of regulatory crimes pay the cost of regulatory failure, which creates inequality and defeats the purpose of the regulation. The British system is cost-effective, but it is not designed to ensure equality and justice consistently; the Italian system is designed to ensure equality and justice more consistently, but it is very costly and legally inefficient (Fioravanti 2011). In both countries, OHS crimes are under-criminalised, but at two different stages of the criminal justice system.

In Britain, Packer's (1968, p. 158) definition of legal efficiency, 'the system's capacity to apprehend, try, convict, and dispose of a high proportion of criminal offenders', seems to be confused with financial efficiency. Legal efficiency, in other words, is not about costs. Enforcement policies in Britain seem to be driven more by cost-efficient than legal efficient decisions (Sanders and Young 2007; Damška 1973, 1986). Again, the literature proposing compliance-driven enforcement policies fails to take into account this issue. Snider (1991) suggests that nation-states using a crime-control model and lacking written and codified constitutions are at risk of witnessing the transformation of the governing classes into political powers serving the interests of the wealthy.

> States will do as little as possible to enforce health and safety laws. They will pass them only when forced to do so by public crisis or union agitation, strengthen them reluctantly, weaken them whenever possible, and enforce them in a manner calculated not to seriously impede profitability. (Snider 1991, pp. 219–220)

Both jurisdictions, however, are affected by global economic pressures. The Italian Constitution and the Constitutional Court responsible for enforcing it are designed to prioritise citizens' civil liberties over money. The guarantees offered by the Italian Constitution, however, are ever more challenged by global neoliberal economic policies and other international institutions and agreements, which constantly aim to decrease state involvement in the economy, cut state's expenditures and its role of guarantor of citizens' civil rights. In other words, while Britain's balance sheet has been gradually adapting to the requirement of a global economic agenda, which has eroded citizens' rights, the opposite is happening in Italy. The opposition that Italian Constitution values are posing to these pressures might also be the cause of the spiralling level of national debt and consequent political crises (Caracciolo 2015; Zagrebelsky 2013). The unregulated nature of the British political system means that the only opposition that a neoliberal economic paradigm can expect is through trade unions' and other social movements' struggles, which, arguably, are also undermined by their ontological convictions to the cause. In Italy, the Constitution represents another layer of opposition, but the subtle, yet monstrous, effect of these changes are threatened by a political class whose decisions are ever more justified by continuous TINA mantras. The increasing under-criminalisation of OHS crimes caused by, in this instance, austerity policies is a symptom of the erosion of civil rights and fundamental modern values of social equality and justice.

References

Ayres, I., & Braithwaite, J. (1992). *Responsive regulation: Transcending the deregulation debate*. New York: Oxford University Press.

Baldwin, R., & Black, J. (2008). Really Responsive Regulation. *The Modern Law Review, 71*(1), 59–94.

Bartrip, P. W. J., & Fenn, P. (1980a). The conventionalization of factory crime a re-assessment. *International Journal of the Sociology of Law, 8*, 175–186.

Bartrip, P. W. J., & Fenn, P. (1980b). The administration of safety: The enforcement policy of the early factory inspectorate 1844–1864. *Public Administration, 58*(Spring), 87–102.

Bartrip, P. W. J., & Fenn, P. (1983). The evolution of regulatory style in the nineteenth century British factory inspectorate. *Journal of Law and Society, 10*, 201–222.

Becker, G. S. (1968). Crime and punishment: An economic approach. *Journal of Political Economy, 76*(2), 169–182.

Caracciolo, L. (2015). *La Costituzione nella palude: Indagine su trattati al di sotto di ogni sospetto.* Imprimatur: Reggio Emilia.

Coffee, J. C. (1981, January). No soul to damn: No body to kick: An unscandalized inquiry into the problem of corporate punishment. *Michigan Law Review, 79*(3), 386–459.

Damaška, M. R. (1973). Evidentiary barriers to conviction and two models of criminal procedure: A comparative study. *University of Pennsylvania Law Review, 506,* 1972–1973.

Damaška, M. R. (1986). *The faces of justice and state authority: A comparative approach to the legal process.* New Haven and London: Yale University Press.

Davis, C. (2004). *Making companies safe: What works?* Centre for Corporate Accountability [Online]. Available from: http://www.unitetheunion.org/uploaded/documents/Making%20Companies%20Safe%20-%20what%20works%20(CCA-Unite%20paper)11-4856.pdf. Accessed 25 Aug 2014.

Fioravanti, M. (2011). Le dottrine dello stato e della costituzione. In R. Romanelli (Ed.), *Storia dello Stato Italiano dall'unità ad Oggi.* Roma: Donzelli.

Fisse, B. (1983, December). Reconstructing corporate criminal law: Deterrence, retribution, fault, and sanctions. *Southern California Law Review, 56*(6), 1141–1246.

Hawkins, K. (1984). *Environment and enforcement: Regulation and the social definition of pollution.* Oxford: Clarendon Press.

Hawkins, K. (2002). *Law as last resort: Prosecution decision-making in a regulatory agency.* Oxford: Oxford University Press.

Hazards Magazine. (2010, October–December). Get shirty. *Hazards Magazine,* issue 112 [Online]. Available from: http://www.hazards.org/votetodie/getshirty.htm. Accessed 25 Aug 2014.

House of Commons Work and Pensions Select Committee. (2004). *The work of the Health and Safety Commission and Executive.* Fourth Report of Session 2003–04, vol. I. HC 456-1. London: The Stationery Office [Online]. Available from: http://www.publications.parliament.uk/pa/cm200304/cmselect/cmworpen/456/45602.htm. Accessed 25 Aug 2014.

HSC. (2002). *Highlights from the HSC annual report and the HSC/E accounts 2001/2002* [Online]. Available from: http://www.hse.gov.uk/aboutus/reports/index.htm. Accessed 25 Aug 2014.

HSC. (2003). *Highlights from the HSC annual report and the HSC/E accounts 2002/2003* [Online]. Available from: http://www.hse.gov.uk/aboutus/reports/index.htm. Accessed 25 Aug 2014.

HSC. (2004). *Highlights from the HSC annual report and the HSC/E accounts 2003/2004* [Online]. Available from: http://www.hse.gov.uk/aboutus/reports/index.htm. Accessed 25 Aug 2014.

HSC. (2005). *Highlights from the HSC annual report and the HSC/E accounts 2004/2005* [Online]. Available from: http://www.hse.gov.uk/aboutus/reports/ index.htm. Accessed 25 Aug 2014.

HSC. (2006). *Highlights from the HSC annual report and the HSC/E accounts 2005/2006* [Online]. Available from: http://www.hse.gov.uk/aboutus/reports/ index.htm. Accessed 25 Aug 2014.

HSC. (2007). *Highlights from the HSC annual report and the HSC/E accounts 2006/2007* [Online]. Available from: http://www.hse.gov.uk/aboutus/reports/ index.htm. Accessed 25 Aug 2014.

HSC. (2008). *Highlights from the HSC annual report and the HSC/E accounts 2007/2008* [Online]. Available from: http://www.hse.gov.uk/aboutus/reports/ index.htm. Accessed 25 Aug 2014.

HSE. (2009). *HSE annual report and accounts 2008/2009* [Online]. Available from: http://www.hse.gov.uk/aboutus/reports/index.htm. Accessed 25 Aug 2014.

HSE. (2010). *HSE annual report and accounts 2009/2010* [Online]. Available from: http://www.hse.gov.uk/aboutus/reports/index.htm. Accessed 25 Aug 2014.

HSE. (2011). *HSE annual report and accounts 2010/2011* [Online]. Available from: http://www.hse.gov.uk/aboutus/reports/index.htm. Accessed 25 Aug 2014.

HSE. (2012). *HSE annual report and accounts 2011/2012* [Online]. Available from: http://www.hse.gov.uk/aboutus/reports/index.htm. Accessed 25 Aug 2014.

HSE. (2013). *HSE annual report and accounts 2012/2013* [Online]. Available from: http://www.hse.gov.uk/aboutus/reports/index.htm. Accessed 25 Aug 2014.

HSE. (2014). *HSE annual report 2013/14* [Online]. Available from: http:// www.hse.gov.uk/aboutus/reports/1314/ar1314.pdf. Accessed 3 Mar 2017.

HSE. (2015). *HSE annual report 2014/15* [Online]. Available from: http:// www.hse.gov.uk/aboutus/reports/ara-2014-15.pdf. Accessed 3 Mar 2017.

HSE. (2016). *HSE annual report 2015/16* [Online]. Available from: https://www. gov.uk/government/uploads/system/uploads/attachment_data/file/534093/ hse-annual-report-and-accounts-2015-2016.pdf. Accessed 3 Mar 2017.

HSE. (2017a). *HSE annual report and accounts 2016/17* [Online]. Available from: http://www.hse.gov.uk/aboutus/reports/ara-2016-17.pdf. Accessed 3 Mar 2017.

HSE. (2017b). *Health and safety at work summary statistics for Great Britain 2017* [Online]. Available from: http://www.hse.gov.uk/statistics/overall/ hssh1617.pdf. Accessed 10 Apr 2017.

INAIL. (2003). *Bilancio Consuntivo 2002* [Online]. Available from: https:// www.inail.it/cs/internet/docs/1_bil-cons-2002-pdf.pdf. Accessed 6 Apr 2018.

INAIL. (2007). *Bilancio Consuntivo 2006* [Online]. Available from: https://www. inail.it/cs/internet/docs/bilancio-cons-2006-pdf.pdf. Accessed 6 Apr 2018.

INAIL. (2014a). *La storia* [Online]. Available from: http://www.inail.it/inter-net/default/Chisiamo/Lastoria/index.html. Accessed 25 Aug 2014.

INAIL. (2014b). *Bilancio Consuntivo 2013* [Online]. Available from: https://www.inail.it/cs/internet/docs/all-bilancio-consuntivo-2013.pdf. Accessed 6 Apr 2018.

INAIL. (2016). *Bilancio Consuntivo 2015* [Online]. Available from: https://www.inail.it/cs/internet/docs/ammt-bilancio-consuntivo-2015.pdf. Accessed 6 Apr 2018.

INAIL. (2017). *Bilancio Previsione 2017* [Online]. Available from: https://www.inail.it/cs/internet/docs/ammt-bilancio-previsione-2017.pdf. Accessed 6 Apr 2018.

Jones, S. (2013). *Criminology*. Oxford: Oxford University Press.

Jost, P. J. (1997). Regulatory enforcement in the presence of a court system. *International Review of Law and Economics, 17,* 491–508.

Ministero della Salute: Direzione generale della programmazione sanitaria, dei livelli di assistenza e dei principi etici di sistema. (2004). *Rapporto nazionale di monitoraggio dei livelli essenziali di assistenza anno 2001* [Online]. Available from: http://www.salute.gov.it/imgs/C_17_pubblicazioni_1175_allegato.pdf. Accessed 25 Aug 2014.

Ministero della Salute: Direzione generale della programmazione sanitaria, dei livelli di assistenza e dei principi etici di sistema. (2006). *Rapporto nazionale di monitoraggio dei livelli essenziali di assistenza anni 2002–2003* [Online]. Available from: http://www.salute.gov.it/imgs/C_17_pubblicazioni_1173_allegato.pdf. Accessed 25 Aug 2014.

Ministero della Salute: Direzione generale della programmazione sanitaria, dei livelli di assistenza e dei principi etici di sistema. (2007). *Rapporto nazionale di monitoraggio dei livelli essenziali di assistenza anno 2004* [Online]. Available from: http://www.salute.gov.it/imgs/C_17_pubblicazioni_1174_allegato.pdf. Accessed 25 Aug 2014.

Ministero della Salute: Direzione generale della programmazione sanitaria, dei livelli di assistenza e dei principi etici di sistema. (2009). *Rapporto nazionale di monitoraggio dei livelli essenziali di assistenza anni 2005–2006* [Online]. Available from: http://www.salute.gov.it/imgs/C_17_pubblicazioni_1072_allegato.pdf. Accessed 25 Aug 2014.

Ministero della Salute: Direzione generale della programmazione sanitaria, dei livelli di assistenza e dei principi etici di sistema. (2010). *Rapporto nazionale di monitoraggio dei livelli essenziali di assistenza anni 2007–2009* [Online]. Available from: http://www.salute.gov.it/imgs/C_17_pubblicazioni_1674_allegato.pdf. Accessed 25 Aug 2014.

Ministry of Housing, Communities & Local Government. (2011). *Annex A8: revenue outturn cultural, environmental and planning services (RO5) 2009–10 (revised)* [Online]. Available from: https://www.gov.uk/government/uploads/system/uploads/attachment_data/file/560117/Annex_A8.xls. Last accessed 3 Apr 2018.

Ministry of Housing, Communities & Local Government. (2012). *Annex A8: revenue outturn cultural, environmental and planning services (RO5) 2010–11* [Online]. Available from: https://www.gov.uk/government/uploads/system/uploads/attachment_data/file/15252/2123447.xls. Last accessed 3 Apr 2018.

Ministry of Housing, Communities & Local Government. (2013). *Annex A8: revenue outturn cultural, environmental, regulatory and planning services (RO5) 2011 to 2012* [Online]. Available from: https://www.gov.uk/government/uploads/system/uploads/attachment_data/file/15356/revenue_outturn_2011-12_final_annex_a8.xls. Last accessed 3 Apr 2018.

Ministry of Housing, Communities & Local Government. (2014). *Annex A8: revenue outturn cultural, environmental, regulatory and planning services (RO5) 2012 to 2013* [Online]. Available from: https://www.gov.uk/government/uploads/system/uploads/attachment_data/file/282502/Annex_A8_Revenue_Outturn_Cultural__Environmental__Regulatory_and_Planning_Services__RO5__2012-13__revised_.xls. Last accessed 3 Apr 2018.

Ministry of Housing, Communities & Local Government. (2015). *Annex A8: revenue outturn cultural, environmental, regulatory and planning services (RO5) 2013 to 2014* [Online]. Available from: https://www.gov.uk/government/uploads/system/uploads/attachment_data/file/379849/Annex_A8_-_Revenue_Outturn_Cultural__Environmental__Regulatory_and_Planning_Services__RO5__2013-14.xlsx. Last accessed 3 Apr 2018.

Ministry of Housing, Communities & Local Government. (2016). *Revenue outturn (RO5) data 2014 to 2015 by local authority (revised)* [Online]. Available from: https://www.gov.uk/government/uploads/system/uploads/attachment_data/file/497101/Revenue_Outturn__RO5__data_2014-15_by_LA_-_02-Feb-2016.xls. Last accessed 3 Apr 2018.

Ministry of Housing, Communities & Local Government. (2017a). *Revenue outturn cultural, environmental, regulatory and planning services (RO5) 2015 to 2016* [Online]. Available from: https://www.gov.uk/government/uploads/system/uploads/attachment_data/file/659793/RO5_2015-16_data_by_LA_-_Revision.xlsx. Last accessed 3 Apr 2018.

Ministry of Housing, Communities & Local Government. (2017b). *Revenue outturn cultural, environmental, regulatory and planning services (RO5) 2016 to 2017* [Online]. Available from: https://www.gov.uk/government/uploads/system/uploads/attachment_data/file/659778/RO5_2016-17_data_by_LA.xlsx. Last accessed 3 Apr 2018.

Nelken, D. (2010). *Comparative criminal justice: Making sense of difference.* London (UK), Thousand Oaks (CA), New Delhi (IND) and Singapore: SAGE Publications Ltd.

Packer, H. L. (1968). *The limits of criminal sanction.* Stanford, CA: Stanford University Press.

Pearce, F. (1976). *Crimes of the powerful: Marxism, crimes and deviance.* London: Pluto Press.

Posner, R. A. (1972). The behaviour of administrative agencies. *The Journal of Legal Studies, 1,* 314–325.

Reiman, J., & Leighton, P. (2010). *The rich get richer and the poor get prison: Ideology class and criminal justice.* Boston: Allyn & Bacon.

Repubblica Italiana. (1994). *Decreto Legislativo n. 626, 19 settembre 1994.* Attuazione delle direttive 89/391/CEE, 89/654/CEE, 89/655/CEE, 89/656/CEE, 90/269/CEE, 90/270/CEE, 90/394/CEE, 90/679/CEE, 93/88/CEE, 95/63/CE, 97/42/CE, 98/24/CE, 99/38/CE, 99/92/CE, 2001/45/CE, 2003/10/CE, 2003/18/CE e 2004/40/CE riguardanti il miglioramento della sicurezza e della salute dei lavoratori durante il lavoro. *Gazzetta Ufficiale* n.265 del 12 novembre 1994. Supplemento Ordinario n. 141.

Sanders, A., & Young, R. (2007). *Criminal justice* (3rd ed.). Oxford: Oxford University Press.

Shavell, S. (1982). The social versus the private incentive to bring suit in a costly legal system. *Journal of Legal Studies, 11,* 333–339.

Shavell, S. (1999). The level of litigation: Private versus social optimality of suit and settlement. *International Review of Law & Economics, 19,* 99–115.

Snider, L. (1991). The regulatory dance: Understanding reform processes in corporate crime. *International Journal of Sociology of Law, 19*(2), 209–237.

Soderberg, M. (2007). Uncertainty and regulatory outcome in the Swedish electricity distribution sector European. *Journal of Law and Economics, 25,* 79–94.

Spitzer, M. (2000). Judicial auditing. *Journal of Legal Studies, 29,* 649–683.

Tinti, B. (2007). *Toghe Rotte.* Roma: Chiarelettere.

Tombs, S., & Whyte, D. (2010). *Regulatory surrender: Death, injury and the non-enforcement of law.* Liverpool: Institute of Employment Rights.

Travaglio, M. (2007). Preface. In B. Tinti, *Toghe Rotte.* Roma: Chiarelettere.

Zagrebelsky, G. (2013). *Fondata sul lavoro. La solitudine dell'articolo 1.* Bologna: Einaudi.

6

Discretionary and Legal Consistency and Proportionality

Another important difference between Britain and Italy is the level of discretion that enforcement officers and institutions can use in their law enforcement activities. As it was observed in the introduction, discretion is defined by the English Oxford Dictionary (2017) as 'the freedom to decide what should be done in a particular situation', which in this context is referred to the decisions of officers and institutions in the course of their professional law enforcement interactions with duty holders. Hence, the use of discretion during regulatory enforcement activities is a highly debated subject because, on the one hand, every interaction requires a level of discretionary decision, but on the other hand, too much discretion can cause discrimination and injustice.

The use of law enforcement discretionary practices is embedded in every level of the British criminal justice system but forbidden in Italy, especially within executive law enforcement agencies. The use of discretion seems a fundamental and yet controversial issue in law enforcement as it gives flexibility to officers and agencies, but it can create inconsistent and unproportionate responses of the criminal justice system to different crimes and social classes. In this context, discretion causes the under-criminalisation of occupational health and safety (OHS) crimes

© The Author(s) 2019   **173**
D. Canciani, *The Politics and Practice of Occupational Health and Safety Law Enforcement*, Critical Criminological Perspectives,
https://doi.org/10.1007/978-3-319-98509-1_6

because these breaches are not treated as other similarly harmful crimes. This is especially the case since equality before the law is a fundamental principle of modern criminal justice systems, but with the use of discretion the law is adapted to specific circumstances, and hence it loses consistency and proportionality and becomes unfair to a section of the population. From this research study, it emerges that discretion should increase gradually and only if the judiciary branch of the state is involved in the decision-making process. Discretion should be forbidden to grassroots law enforcement officers.

This topic emerged very often during the interviews with British and Italian enforcement officers and it has been a long-standing debate among academics and policymakers. Thus, the findings on this topic have been organised into two chapters. This chapter concentrates on the enforcement officers' opinions of the relationship between discretion and the principle of legal consistency, proportionality and fairness. The following (Chapter 7) analyses, compares and discusses how discretionary practices affect the work and decisions that OHS enforcement officers take during their operations.

The use of discretion during enforcement activities is important because enforcement officers' and institution's decisions to adopt compliance-driven or deterrence-driven enforcement policies are influenced by the context in which the regulatory breach occurs. The use of compliance-driven enforcement policies prevents the enforcement institutions from becoming over-reliant on strict, unreasonable and unfair legal requirements (Bardach and Kagan 1982). Hence, the use of discretion is analysed in relation to the everyday work of inspectors (Braithwaite 1987; Hawkins 1984; Hutter 1997; Black 2001), the advantageous effects it might have in terms of achieving compliance, such as gaining the trust of duty holders (Dworkin 1977; Galligan 1986), but also on the disadvantageous effects it might have on the enforcement institution and criminal justice system (Richardson et al. 1983).

Hawkins (1984), who conducted the only available major research study on British OHS enforcement activities, supports the use of discretion during enforcement activities. His analysis of discretionary practices during legal enforcement, however, is based on the 'interpretative sociology in the tradition of the societal reaction school' (Hawkins 1984, p. 15) of Howard Becker (1963, 1964) and 'the dramaturgical approach' (Hawkins 1984, p. 15) of Erving Goffman (1959, 1961, 1963, 1967, 1970, 1971). Hawkins (1984) argues that his theory on the use of discretion during enforcement is based on 'ethnographic analyses of discretionary behaviour in legal settings' (Hawkins 1984, p. 15), such as Cicourel (1968), Emerson (1969), Manning (1977), Reiss (1971), Ross (1970), Skolnick (1966), and Sudnow (1965). Hence, in the context of the European goal-setting OHS regulations, where requirements are not prescriptive or clearly defined, discretion becomes an essential tool to achieve compliance efficiently (Baldwin and Hawkins 1984; Black 2001). Discretion allows enforcement officers and duty holders to negotiate regulatory compliance (Meindinger 1986; Olsen 1992). When analysed in these terms, discretion is an indispensable tool for enforcement officers and institutions, but its benefits are questioned when analysed from a different perspective. Richardson et al. (1983) and Galligan (1986) argue that the decision to allow discretion raises a familiar dilemma between crime prevention and just desserts as sentencing aims, because adapting compliance levels to duty holders' needs does not ensure a consistent application of the law between citizens, and raises fundamental questions about the achievement of the equality before the law principle of the modern legal systems. In other words, the fundamental question to ask is, whom should the law be fair to?

These legal principles can differ radically between the common law and civil law systems and, in particular, between jurisdictions where criminal justice system policies have traditionally been based on due-process or crimes-control principles. While the former system is based on the idea of respecting *written statutes* and *institutions*, the latter is based on the application of 'discrete inductive ideas capable of functioning only within limited factual spheres' (Legrand 1996, p. 65), which are not necessarily rational or logical but that create forms of common social and, therefore, legitimate, understanding (Legrand 1996). This is a significant difference between the British legal system, which is traditionally based on crime-control principles, and the Italian, which is based on due-process ones. While scholars advocating a discretionary compliance-driven approach to enforcement highlight the advantages of adapting the law to reach an acceptable level of compliance, those advocating a deterrence-driven approach highlight the advantages of ensuring the basic legal principle of equality before the law, which is achieved through the use of consistent and proportionate legal enforcement actions across the criminal justice system spectrum (Richardson et al. 1983; Galligan 1986). In other words, the under-criminalisation of OHS crimes occur because the reaction of the criminal justice system to the harm caused by these crimes is *inconsistent* with other similarly harmful crimes.

The use of discretion among criminal justice system institutions indicates that the relationship between a state's institutions and people is not strictly regulated. As was mentioned in the introduction, the separation of powers doctrine is a fundamental aspect of this relationship because this principle regulates the relationships between the three main branches of the state in order to preserve individual freedoms and rights. Montesquieu (de Secondat 2001) argues that to avoid despotic rulers, guarantee personal civil liberties and create democratic governments, the state should be formed of three main institutions, each with specific responsibilities: legislating (legislative), executing legislations (executive) and judging legal violations (judiciary). Montesquieu's theory was inspired by Roman law and British constitutional law, which were both based on the tripartite system (Sabine and Thorson 1973). Yet the British legislature and executive had, and still has, a political relationship between institutions that is much more intertwined and less regulated than the one envisaged by Montesquieu. This is a major difference between Britain and Italy and a fundamental difference between the common law and statutes law traditions (Vile 1963; Langbein et al. 2009; Mousourakis 2015). Hence, to ensure civil liberties, equality, democratic principles and avoid civil war, these three institutions should be mutually dependent on each other, but also scrutinise each other's operations through processes of checks and balances.

The extent of discretionary decision of law enforcement institutions, as well as the level of independence they have from the government, parliament or people are fundamental aspects to consider in this context. Hence, the way the separation of powers doctrine is applied in practice in these two jurisdictions is fundamental in this context. As it was observed in the previous chapter, resources can have significant impact on enforcement policies. On the one hand, the way the separation of powers doctrine and the principle of legal consistency is interpreted and applied in Italy means that the government cannot decide the resources to be given to the institution because these might affect their operations and breach fundamental Italian legal and social principles, which are declared in the Italian constitution. On the other case, this is not the case in Britain, where the government, through the Secretary of State for Work and Pensions (SoSWP), has a significant influence over the budget of the OHS enforcement institution and the enforcement policies they adopt. The British Parliament has limited power of decision on this issue.

In terms of enforcement policies, this is also different. On the one hand, in Italy, neither the government nor the OHS enforcement institution can decide the enforcement policies to use. For the sake of legal consistency and proportionality, this is strictly regulated by the Code of Penal Procedure (CPP) and the Judiciary Police Official (UPG) regulation. On the other hand, the Health and Safety Executive (HSE) enforcement decisions are based on the organisation Enforcement Policy Statement and the Enforcement Management Model (EMM) (HSE 2015), which are agreed between various criminal justice system institutions, including the Crown Prosecution Service (CPS), but these are not nationwide strategies such as the CPP in Italy.

The Italian OHS enforcement institution is also forbidden from choosing the legal cases that should be tried in court and the Public Ministry (PM) is obliged to initiate prosecution trials if there is enough evidence of criminal conduct. That responsibility belongs to public prosecutors, who are members of the PM, which is a branch of the state judiciary (Gustapane 2012), and thus independent from the government and parliament. In Britain, the CPS has similar functions to the PM in Italy, but not in the case of OHS crimes. The OHS enforcement institutions retain the power to decides which cases should be tried in court and also the responsibility to directly represent the claimant's interests (HSE 2015). Indeed, this is regulated by the EMM and Enforcement Policy Statement, but in the previous chapter it was demonstrated that these decisions are also taken according to the resources available.

The law enforcement institutions' freedom to decide triable cases was the subject of lengthy debates in Britain. Since 1981, there have been questions raised in relation to allowing enforcement institutions, such as the British Police, the discretion to decide whether a case is triable. The Law Commission proposed, in the 1981 *Royal Commission on Criminal Justice* (a.k.a. Philips) report, to remove the responsibility of enforcement institutions to proceed with prosecutions (either themselves or through selecting prosecutors from private law firms). The report led to the institution of the CPS, which aimed to create more legal consistency across criminal justice system institutions, crimes and geographical areas in Britain.[1] The institution of the CPS is a typical example where the executive powers of the government, which controlled the British Police, were restricted by additional checks and balances aiming at avoiding abuses of powers and inconsistent and unproportionate legal responses to crimes. The institution of the CPS, thus, suggests that legal inconsistency has been recognised as a problem in the British legal system.[2] The HSE, however, retains almost all its prosecutorial discretionary powers, but has increasingly implemented enforcement (procedural) models (EMM) and agreed to CPS-approved guidelines and protocols to repair this issue.

However, the under-criminalisation of OHS crime is proof that discrepancies still exist. The HSE enforcement policy statement tries to standardise enforcement responses. Besides the fact that the statement can easily be amended, it also clearly states that enforcement decisions depend, also, on resources available to the institution, or subject to the public interest test. The CPS's aim of achieving more legal consistency across the British criminal justice system was criticised in a 2009 House of Common Justice Committee (2009) report, which concluded that

---

[1]According to Damaška (1973), the establishment of prosecution services, such as the Crown Prosecution Service, in jurisdictions adopting an adversarial legal system is a recent phenomenon. In Scotland this responsibility is given to the Procurator Fiscal Service. See also footnote 2 in the second chapter (Histories and Traditions). It is important to note that these models are different in *principle* and that a lengthier and more in-depth analysis of these two jurisdictions' criminal justice system policies might offer a richer account of their differences. Unfortunately, a lengthier and deeper analysis of the two jurisdictions' criminal justice systems is beyond the primary scope of this research study.

[2]More in England and Wales, as compared to Scotland.

Inconsistency in Crown Prosecution Service delivery was a clear theme in the evidence […] Failures to define clearly the role of the prosecutor, and the pressures pushing and pulling it in different directions, militate against priorities for consistent delivery [and raises] questions about what kind of local discretion is desirable and beneficial to the public interest. The Attorney General should make a clear statement of how local responsiveness can be made compatible with the demands of natural justice for system-wide consistency. (House of Common Justice Committee 2009, par. 114)

The justification to avoid the transfer of the HSE prosecution decisions to the CPS is due to the concern that centralising judicial decisions might diminish the public interest involvement in local enforcement and judiciary activities. The current arrangement is also based on the assumption that the HSE is the best institution to understand the needs of citizens through processes of accountability. However, the government retains great control of its operations, the HSE shows little accountability to the public, and the enforcement policies adopted are not consistent when compared to other non-OHS legal breaches.

In Italy, the criminal justice system is much stricter in terms of the application of the separation of powers doctrine and complies much more strictly to due-process model legal principles (Fioravanti 2011).[3] One of the issues around granting discretion to front-line enforcement officers is that executive interests would clash with judiciary interests and cause under- or over-criminalisation of selected deviant behaviours when compared to other similarly harmful ones. In other words, according to Italian OHS enforcement policies, enforcement officers should only be given the power to *report* crimes to public prosecutors operating within the PM, who decide whether to *apply* the law based on the Penal Code (CP), the CPP principles and the criminal evidence available. These checks and balances eliminate possible conflicts of interest that would alter the criminalisation of specific social activities or groups. Hence, the PM is a Judiciary Authority institution independent

---

[3]The 1948 Italian Constitution employed the institutional separation principle because the post-war Italian political class wanted to ensure the preservation of individual civil liberties that were denied during Mussolini's dictatorship (Fioravanti 2011).

from the government, and it controls the OHS enforcement officers at arm's length. This system is designed to ensure the due-process principles of legal proportionality and consistency across the Italian criminal justice system. It also consents the use of a certain degree of discretion, but only through a close supervision by the PM. Prohibiting the use of personal discretion to front-line Prevention Services for Safety in Work Environments's (SPSAL) enforcement officers was essential in the 1994 regulatory and enforcement policies reforms. In 1994, officers acquired UPG powers in order to conform their activities and enforcement behaviours to that of other law enforcement officers, who requires them to follow the CPP and operate at arm's length from public prosecutors (the judiciary). Hence, in contrast to the British system, where the use of discretion is not strictly regulated, the use of discretion in Italy is forbidden among enforcement institutions and officers, and strictly regulated between the executive, legislative and judiciary.

The use of discretion is also significant in the use of compliance-driven enforcement policies. Richardson et al. (1983) argue that the discretion enjoyed by enforcement officers might be part of an invisible problem. Enforcement officers need to place a great level of trust on duty holders in order to ensure compliance effectively (Braithwaite 1997; Hawkins 2002), which can create regulatory capture and result in inconsistent and disproportionate enforcement decisions. In other words, compliance-driven practices coupled with discretionary powers can create capture and unintentional decriminalisation of crimes (Carson 1970a, 1979; Fooks 2008; CCA 2008). Discretionary judgements might be influenced by social, economic and political local, national and global trends, such as funding, governments' agendas, levels of unemployment, or the media's and public's attitudes towards specific issues.

An advantage of relying on discretionary practices is that the institutions are quick to respond to sudden regulatory issues and react effectively and efficiently to emergency situations (Hawkins 2002). The disadvantage, however, is that the organisation might find justifications for implementing policies that are against citizens' interests, as happened following the 2000 *Revitalising health and safety* consultation (Davis 2004). Increasing transparency or the level of the organisation's accountability to the public might be a solution, as is the case of the British Police (Newburn and Reiner 2007). However, as it was observed in Chapter 4, the HSE is not keen to be transparent and the level of accountability it has towards citizens cannot be compared to that of the British Police. Untested assumptions about the regulation enforced and the contemporary socio-economic context have played a considerable role in shaping the individual officers' and institutions' reactions to legal violations. In fact, Carson (1970a), Bartrip and Fenn (1980a, 1980b), and Gunningham and Johnstone (1999) all suggest that the chronic under-criminalisation of safety crimes has been caused by discretionary enforcement decisions affected by inadequate political, judicial and executive responses. Since the early 2000s, various political rhetoric has continually criticised regulations as burden to the economy, and given to the government the legitimacy to deprive the HSE of its fundamental financial funding needed to maintain a minimum level of deterrence-driven enforcement activities. Yet, safety incidents *and* occupational sicknesses remain one of the major causes of premature death in developed countries (Tombs and Whyte 2007, 2009, 2010).

This chapter and Chapter 7 come in tandem as they both scrutinise how discretion affects OHS enforcement policies in Britain and in Italy. This chapter is more concerned with the British and Italian enforcement officers' opinions on the relationship between the use of discretion and the importance of achieving fundamental modern criminal justice system principles of legal consistency, proportionality and fairness. Chapter 7 analyses, compares and discusses how discretionary practices affect the work and decisions that OHS enforcement officers take during their operations in Britain and Italy.

1 Discretion, Legal Consistency, Proportionality and Fairness

This section focuses on the British and Italian officers' understanding of the use of discretion during enforcement activities and how it is affected by key legal principles of legal consistency, proportionality and fairness. British and Italian OHS enforcement officers showed awareness of these

principles. The principles of legal fairness, consistency and proportionality endorsed by modern criminal justice systems ensure that legal litigations are not affected by biases and impartialities and that policing
institutions' responses are fair and, hence, consistent and proportionate
when interacting with citizens.

This section demonstrates that the under-criminalisation of OHS
crimes in Britain is a result of an inconsistent and unproportionate
response to these crimes caused by discretionary practices. While the
under-criminalisation of OHS crimes is an issue also in Italy, this is not
occurring at field enforcement level, as is the case in Britain. During
interviews, enforcement officers from both jurisdictions mentioned a
complex interrelationship of enforcement policies and codes and practices, which are all there to ensure that enforcement officers achieve fair
decisions.

This section is divided into two parts. The first part analyses how the
use of discretion among British OHS enforcement officers and institution during enforcement causes inconsistent and unproportionate
responses. The second part of this section critically analyses the Italian
enforcement officers' responses to the topic and what changed in 1994
when they lost the discretionary enforcement powers they had enjoyed
since their foundation in 1980.

1.1   Britain

Health and safety enforcement officers considered consistency of
enforcement actions to be the achievement of a uniform level of regulatory compliance across the business community. The EMM
helps officers to ensure enforcement consistency across duty holders.
However, enforcement officers defined consistency as the achievement
of the same outcomes rather than ensuring equal responses for similar breaches across the whole criminal law spectrum. This was demonstrated when a British enforcement officers argued that

> In terms of consistency we don't say we will do the same thing every time.
> We say broadly that we would seek to achieve the same outcome. (British
> interview 2, Line 461)

This statement indicates that compliance-driven policies are important to prevent incidents and crimes, but responses fail to achieve consistency and proportionality. Another British enforcement officer argued that

> The expectation is that the inspectors would act in a consistent way. That doesn't mean that they have to actually act in a uniform way, because there have to be various factors taken into account in reaching the decision. The EMM structures the way they do that, but it doesn't affect the inspector by saying that you have to do that, that you have to prosecute for that, or you don't have to prosecute for that, but it does give a structure to what we do, which is to ensure the perception of fairness. (British interview 3, Line 744)

This officer also differentiated between legal consistency and uniformity. From the interviews it emerged that the former requires the enforcement officers to act similarly when finding similar regulatory breaches, while the latter requires the enforcement officers to achieve similar regulatory outcomes between firms. British enforcement policies give much more importance to the achievement of uniform regulatory outcomes, rather than ensuring that the enforcement officers react consistently while inspecting firms. They argued that uniformity is achieved by using the EMM, which standardises enforcement officers' behaviours and decisions relating to breaches and allows officers to take into account characteristics—such as employers' willingness to comply—to achieve regulatory compliance and workers' safety. Hence, enforcement officers' behaviours are not consistent in the broader legal definition of the term. Instead, their reactions are uniform because they try to achieve the same regulatory outcomes. This is a problem because, as a result, employers, especially those showing a willingness to cooperate, are positively discriminated when compared to other crimes and, therefore, officers' reactions contribute to the under-criminalisation of OHS crimes. The same enforcement officer continued by arguing that

> [If] we find a couple of things that are actually wrong, but a lot of things that are right and a good system in place, which appears to be working most of the time, we have the discretion to take a lenient approach on the things that we found wrong. (British interview 3, Line 754)

This is evidence that enforcement officers judge and assess specific OHS situations in the work environment and take decisions with regard to the best means to achieve an optimal level of compliance. However, in this example, the officer does not connect the use of a lenient approach to inconsistent enforcement of the law. Various factors are assessed, such as the firm's administrative arrangements and duty holders' attitudes, but these decisions also rely on personal judgements which might vary greatly among officers. However, a British enforcement officer recognised this issue when noting that

> It does call for fairly advanced judgement skills on the part of the inspectors. To form an accurate perception, you know? Is what you see the truth perception, or is it just a story that the management is trying to convince you on the day? Obviously, as human beings we don't always get it right. (British interview 3, Line 772)

This quotation is evidence that enforcement officers might unintentionally take wrong decisions, because duty holders may lie and hide or withhold important information during inspections. The interviewee argued, however, that enforcement officers should be capable of detecting lies and deceptions by making sure that the attitudes shown by firms are supported by concrete actions. In other words, enforcement officers with a fair amount of experience take into consideration the possibility that employers try to deceive them to gain inspectors' trust and a less punitive enforcement response.

The advantages of adjusting regulatory behaviours to specific situations are that employers will always feel as if the enforcement institutions' response is fair towards them, and therefore they will be more prone to listen and follow officers' instructions, which is an idea that was frequently shared among British enforcement officers. There are disadvantages, however. Firstly, enforcement officers' judgements depend on their constructed ideas of compliance, rather than the legal definition. Hence, this confirms Braithwaite's (1985, 1987) idea that enforcement institutions' and duty holders' cooperation is essential to construct the ideal level of compliance expected. Yet there is no mention in the literature that duty holders might also lie and deceive the enforcement institution. Weather discretion is used or not, the issue is that enforcement officers do not have access to the information needed to do their job and must, thus, rely on duty holders' cooperation. A regulatory framework requiring firms to make available important records and documents, as is the case for the OHS risk assessment, would allow enforcement officers to access essential information without the need to cooperate with the duty holders.

Second, implementing regulations in firms can be a very complex task, let alone enforcing them. Bardach and Kagan (1982) and Hawkins (2002) argument that a punitive approach might lead to an unreasonableness application of the law is sensible, but this flexibility is an issue, especially in Britain. Legal consistency should achieve similar responses, not uniform outcomes and the idea of reasonability might develop during interactions between enforcement officers and duty holders. In other words, it might lead to regulatory capture, which was another issue mentioned by the British enforcement officers. The HSE has been criticised for the lack of commitment to deterrence-driven enforcement policies (Tombs and Whyte 2007). Also, Pearce (1983), Baldwin and Veljanovski (1984) and Posner (2003) argue that inadequate enforcement actions caused by the use of discretion, and together with a lack of resources leads to under-enforcement, under-criminalisation, market deregulation, and the erosion of workers' rights.

Thirdly, the EMM helps enforcement officers to reach consistent outcomes when enforcing the law in firms. There are numerous steps to follow during enforcement and, as one of the interviewees argues, 'you would never have a case whereby an inspector makes a decision, acts on that, and prosecutes a business in isolation' (British interview 6, Line 369). Yet, the application of the EMM also depends on the level of resources available, which have significantly affected enforcement policies in the past (Bartrip and Fenn 1980a, b, 1983; Tombs and Whyte 2010). Therefore, the EMM is used as a framework to aid enforcement officers' decisions while devising ways to help firms achieve regulatory compliance, but this means that their behaviour while inspecting is not legally consistent with other breaches, and hence creates unequal responses and injustice. In addition, although the enforcement officers are never in isolation when taking enforcement decisions, it does not mean that the collective decisions undertaken within the HSE and local authorities (LAs) ensure a consistent enforcement of the law. The culture and tradition of the organisation might be at the basis of the under-criminalisation of OHS crimes.

A British enforcement officer argued that some officers might not enforce the law sufficiently

> When I was a middle partnership officer, one of my tasks was to look through the enforcement activities for every local authority that I was responsible for liaising with, and some of them have not served notices in a whole year. For me that was shocking. I cannot believe that in over a year, in a district, there is not one thing that requires a notice. (British interview 8, Line 351)

This is an example that in certain geographical areas the regulation might be intentionally under-enforced. Regulatory capture might cause under-enforcement, but this is only speculation as it is difficult to know the rationale behind individual enforcement officers' decisions. In fact, the blame might not be placed only on enforcement officers' malicious intentions. The same officer also

> worked in a very rural authority before moving to a real urban authority. And in the rural authority I felt very much on my own and I wasn't very experienced and there was nobody to ask. And the legal team was very inexperienced as well, because prosecutions were not being taken. That, to me, was a hindrance because I was lacking in confidence, I didn't know what to do and instead of ringing colleagues in the authorities next door and saying 'um, do you mind giving me some guidance' […] you know, that feels horrible. (British interview 8, Line 173)

Therefore, inconsistent responses to OHS crimes might also be caused by inappropriate training or lack of institutional support and guidance. This demonstrates that empowering enforcement officers to use discretion increased the opportunities to enforce the law inconsistently, whether it be due to lack of resources and support, officers' personal discretionary reasons or top-down institutional and political reasons. Indeed, from the interview it emerged that allowing discretion might cause law enforcement officers and institutions to become overzealous and undermine citizens' civil rights, but the records so far have shown that in the case of OHS the opposite is happening. There are clear indications that discretion plays an important role in the under-criminalisation of OHS crimes in Britain because it allows enforcement officers to turn a blind eye, intentionally or unintentionally, on regulatory misconducts.

If enforcement activities are inconsistent, they might also be un-proportionate. From the interviews, it emerged that OHS enforcement institutions seem to be supportive of employers' needs, and reluctant to impose sanctions on wrong-doers, as the literature arguing for the benefit of using compliance-driven approach confirms (Braithwaite 1987; Hawkins 1984; Hutter 1997; Black 2001). Although similar OHS crimes might attract proportionate responses, when these responses are compared to other crimes this is not the case (Carson 1970a, 1979; Fooks 2008; CCA 2008).

With regard to the issue of consistency, the OHS officers interviewed seem to define legal proportionality differently to their Italian colleagues. While talking about the enforcement officers' activities, a British enforcement officer argued that the use of proportionality for similar breaches rather than to achieve similar outcomes during enforcement

> would change their credibility [...]. The HSE has had huge support from businesses in the past. I think that the large part of that is because inspectors use discretion to have a proportional response. So, we are not seen as overly nit-picking. (British interview 7, Line 291)

Again, legal proportionality was not defined as enforcement and legal proportionality across firms and other crimes. In other words, in Britain, enforcement officers' responses do not seem to be proportionate. So this definition does not consider the broader ideas of social justice and equality before the law underpinning these principles and inspired by modern liberal political values. Proportional response is, indeed, meant from the duty holder's viewpoint, not from its wider legal meaning. When prompted about the use of proportionate responses across criminal breaches, the British officer continued by arguing that if a business does not perceive it [the enforcement action] as proportionate, then HSE loses credibility, and when you try giving information it doesn't work, people will ignore it. [...] So, if you undermine health and safety as a brand, by making it too stringent and if it doesn't actually fit with duty holders' and workers' own perceptions of what the risks are, then you devalue it. (British interview 7, Line 298)

Comparing the law to the same credibility of a 'brand' might demonstrate that the OHS regulation in Britain is already devaluated. Regulatory capture might be another issue here. The officer, in this case, demonstrates that accommodating duty holders' perceptions is more important than ensuring the equality before the law principle. Most duty holders probably comply with officers' instructions, especially if improvement and prohibition notices are issued. Until 2012, however, there were no significant consequences for duty holders when found in breach of most OHS regulations. It was only in 2012 that the HSE empowered to charge employers for their interventions (through the FFI), such as the issuing on-the-spot penalties, which aim to encourage duty holders to achieve compliance before inspections rather than after. Hence, allowing enforcement officers and their institution to use discretion during inspections can lead to a voluntary or involuntary form of capture. When asked whether this approach might consist of a form of regulatory capture, the same officer argued that

it's possible. I think that one of the ways that has always been within the HSE to avoid that is that you get rotated. So, you move around, you start doing different industries after a while. There's always been an accusation that there is regulatory capture, but it's a fine balance because if you don't know the industry, then you lose credibility. So, it is hard work. I think that sometimes you just have to... you have to trust the professional discretion of the inspectors [...] I think that regulatory capture is more of a risk not for whole industries, but an individual company. (British interview 7, Line 336)

Arguably the primary aim of the enforcement institution, officers and the state should be to maintain the credibility of the law and the criminal justice system, rather than the credibility of the industry towards them. This demonstrates that HSE and LA enforcement actions and decisions are captured. Yet, this allegation was denied by the same interviewee, who continued arguing that firms become HSE resistant if more punitive actions are adopted. According to Bartrip and Fenn (1980a) the need to build enforcement institution's credibility towards duty holders started in the mid-nineteenth century, due to the incapability of the Factory Inspectorate to take firms to court and the limited resources available. In fact, their Italian colleague, it will be observed later, did not think that building credibility with employers is important. Therefore, without a proportionate—or sensible—use of the law, firms would boycott OHS unless directly threatened by detection.

The British OHS enforcement institutions adopt mainly compliance-driven approaches during enforcement, but without recognising that the accommodation of industries' needs is a form of regulatory capture. Hence, discretion is used to achieve enforcers' proportionate responses, but not when compared to other non-OHS crimes (Carson 1970a, 1979; Fooks 2008; CCA 2008). Some interviewees recognised this issue only after prompts. A LA officer described working with other non-OHS regulatory agencies as shocking, due to the enforcement officers' hostile attitudes towards employers (British interview 8). Also, the force used by police officers to arrest an employer for corporate manslaughter witnessed by one interviewee seemed to be above reasonableness (British interview 4). All these approaches might be the result of 170 years of OHS enforcement traditions and attitudes, which have been considered to be efficient in monetary terms and to achieve compliance, but have definitely represented a shortcoming in justice terms.

Another aspect that emerged from the interviews while talking about discretion was the officers' perception of fair enforcement actions. Officers' actions might aim to ensure fairness for workers, victims or employers, but given the under-criminalisation of OHS crimes, the expectation is that the fairness mentioned by officers during interviews was referring mostly to employers. Indeed, working on employers' compliance reduces incidents and victims, but a fundamental argument of this research study and the literature is whether OHS regulations should be enforced with deterrence- or compliance-driven policies. Therefore, enforcement officers' discretion creates fair responses towards OHS duty holders, but creates unfairness in the criminal justice system response in Britain.

Ensuring fairness, in addition, might not depend on individual enforcement officers, but on institutional practices and policies, and officers might have little power to change them. This is demonstrated by the fact that (a) British enforcement officers have high vocational commitment to improve OHS; (b) no enforcement officer interviewed could be certain that compliance-driven policies are the most effective in improving standards; (c) it is documented that deterrence-driven policies were adopted for a short period in the nineteenth century only (see Bartrip and Fenn 1980b); (d) reducing levels of resources of British enforcement institutions impair their capability to implement more deterrence-driven policies; and (e) according to the interviewees, deterrence-driven actions might weaken the credibility of the HSE as a whole. Hence, the idea of fairness in the British OHS enforcement institution means that it is right to be punitive towards businesses, but only as a last resort, only when the resources allow it, and only if it is believed to be a fair response. This is a central argument of Hawkins's (2002) support for compliance-driven enforcement policies, but, arguably, a major reason for the under-criminalisation of OHS crimes in Britain.

According to one of the interviewees, if the HSE were to adopt a more deterrence-driven enforcement policy and use less discretion, it might lose credibility, and businesses might start boycotting the law intentionally. With inappropriate levels of resources, the HSE might not be able to enforce the legislation, leading to a deterioration of safety standards. Hence, it can be concluded that the HSE needs to use discretion and appear fair to businesses because with the current level of resources and political support it is simply unfeasible to achieve compliance with deterrence-driven policies.

Therefore, the discretionary policies used by the enforcement institutions in Britain might be fair to businesses, but unfair to workers, OHS victims and other citizens. This research study also proves that the OHS enforcement institutions are captured, because enforcement is based on cooperation with industries occurring through mediating practices, rather than the imposition of strict legal duties. Enforcement institutions might be incapable of adopting deterrence-driven enforcement policies effectively by, for example, taking more cases to court. Enforcement fairness, therefore, is a technique used to achieve some enforcement results at low cost and the British OHS enforcement officers' interpretation of the concepts of legal consistency, proportionality and fairness differ from their original legal meanings.

1.2   Italy

In Italy, the separation of powers doctrine is taken seriously by the enforcement officers and institution. Contrary to the British officers, when the Italian inspectors referred to the separation of powers doctrine, and the principle of ensuring legal consistency and proportionality, they immediately referred to the UPG's duties. The UPG duties, however, were also used as a defensive shield to avoid critics. In fact, the unsupervised nature of the officer's job means that, as will be observed below, officers still use a degree of discretion in their day-to-day operations. Officers are required to follow UPG enforcement rules, but they also considered these obligations an obstacle to promote regulatory compliance effectively among duty holders. Another aspect that was observed during the Italian interviews was that they never referred to the legal principle of fairness. That is because ensuring consistency and proportionality already represents a fair response for employers, but also workers and other citizens. Hence, the British officers' definition of enforcement fairness does not apply in Italy because Italian officers must be fair towards everyone by law, not just OHS duty holders. A Labour Territorial Inspectorates (ITL) officer argued that

> The inspectors have institutional responsibilities obliging them to behave as Judiciary Police Officials [UPG]. It is not possible to use discretion on the application of the regulation. The job of the inspectors is not doing justice, but to refer the facts of penal relevance detected to the Judicial Authority, which then decides if to send the employer to judgement. (Italian interview 10, Line 23)

This is the enforcement officer's mandate explained by law, and embraces the due-process principle of criminal justice, which is used to ensure constant and proportionate responses. It might be possible that explaining enforcement duties by the law allows them to avoid talking about how enforcement *really* occurs. After all, the enforcement officers just have to report breaches, which means that the consequences of OHS crimes trials do not concern them as much as it does their British colleagues. For Italian OHS enforcement officers the prosecution of OHS crimes is dealt with by public prosecutors within the PM, and besides their responsibility to collect evidence during investigation and their appearance in court as witnesses, they are not involved.

When talking about the issue of low levels of penalties and the officers' incapability of controlling this issue, an Italian enforcement officer argued that

> There was a temptation to empower SPSAL to decide which enforcement policy to use for administrative breaches. By doing this, the risk was that the mediation process [between the enforcement institution and businesses] would fall entirely within local SPSAL. This system would require the involvement of different managerial levels within the SPSAL when deciding the administrative penalties to apply to non-compliant firms. In terms of administrative breaches, this system could even lead to the elimination of administrative penalties for enforcement purposes. But since the administrative justice system in Italy has completely collapsed, it is difficult to imagine moving such a delicate aspect of the Italian enforcement policies within the SPSAL. After all, the current enforcement policies are effective and must be applied this way. (Italian interview 9, Line 401)

A few interesting observations can be made about this statement. Firstly, during the interview this officer was supportive of regaining more enforcement discretionary powers. They enjoyed a similar level of discretion to that of the British officers until 1994, when they were also not obliged to issue penalties and, hence, enforcement practices were more compliance-driven. Secondly, the officer argued that if the SPSAL were given more discretionary powers, their enforcement policies might become more lenient. Thirdly, the officer also argued that the current deterrence-driven policies are effective for achieving compliance. Fourthly, reform conferring more discretionary powers to OHS enforcement officers would have been appreciated by this officer, but it was also perceived as problematic because it would create a system that is more difficult to manage. The system would increase administration workload and costs, but it might also undermine the consistent and proportionate response of the Italian criminal justice system. The officer also argued that because the administration of the criminal justice system in Italy has collapsed, devolving SPSAL's enforcement powers would be almost unimaginable, which depicts both the complexity the reformation process would entail, as well as providing a clear picture of the condition of the Italian Judiciary Authority.

Several more enforcement officers complained that they do not have enough discretionary powers. For this reason, there is also an element of conflict between the enforcement institution and the Judicial Authority. These clashes occur because the enforcement officers' obligations to report every criminal breach to public prosecutors runs counter to the 'working methods that belongs to the culture of the institution [SPSAL]' (Italian interview 1, Line 410). Hence, an Italian enforcement officer argued that

> It's okay surveillance, but there is a very heavy, pressing obligation on reporting to the Judiciary all of those deficiencies, hence, crimes. (Italian interview 3, Line 69)

It is worth noting that in the first quote the interviewee mentioned the word 'culture' to explain the traditional discretionary and compliance-driven practices employed by SPSALs from their foundation in 1980 up until 1994. They were also allowed to offer free-of-charge technical advice, as private consultants do, to employers. All this changed in 1994 when the enforcement institution started adopting a more deterrence-driven enforcement approach in order to comply with UPG regulation and CPP. It is also interesting to note how terms such as 'heavy' and 'pressing' used by the officer in the second quote indicates how officers perceive the obligation imposed by these laws.

These two quotes depict an internal conflict between the primary perceived prevention aim of SPSAL officers and the function imposed on them by the UPG regulation and the CPP. It is interesting because when decentralised, discretion confers more power to the enforcement institutions, but its primary cultural and traditional objective might run counter to the criminal justice system's objective of achieving the equality before the law principle. Thus, discretionary practices empower the enforcement institutions and officers, but at a price. They know best how to achieve compliance, but when given discretionary power they devaluate the importance of achieving equality of justice before the law. This, in Britain, under-criminalises OHS crimes. Italian enforcement institutions' and officers' discretionary practices are very restricted because the equality before the law principle is as important as achieving compliance, but such practices are much more ineffective and costly.

However, these law enforcement practices were introduced to conform to UPG and CPP enforcement principles as used by other law enforcement agencies.

> So, the Ministry of Health says: 'prevention, prevention'. The region says: 'prevention'. We are saying prevention. But then I tell them: 'But what about the magistrates [public prosecutor]?' And they reply: 'It is absolutely necessary that the law is respected!' (Italian interview 5, Line 386)

This quotation depicts well how different institutions would like SPSAL officers to use more prevention, or lenient compliance-driven enforcement policies, but judiciary-imposed practices forbid these. This conflictual relationship between executive and judiciary institutional interests is a central tenant of the checks and balances of the separation of powers doctrine. In Italy, the enforcement institutions' prohibition to use discretion and the sole PM public prosecutor's responsibility to decide whether to start prosecution ensure that separation. Given this division, questioning which enforcement approach officers felt more attuned with led to very interesting responses.

> I am part of the health service, and I never forget this. Besides, I say to the Judiciary: I am part of the health service and I have received these instructions. On the other hand, if the Judicial Authority recognises that we are not [i.e. we follow UPG rules]… I become like a Carabiniere. But I am not a Carabiniere. […] It is difficult to synchronise with the Judicial Authority because they talk at another level. (5th Italian interview 5, Line 390)

This is yet further evidence that there are conflicts between SPSAL officers' enforcement traditions, who would like to use more compliance-driven enforcement policies, and the Judiciary Authority, which does not want enforcement officers to breach UPG principles, and hence, to use compliance-driven policies. Another aspect argued by the interviewee was that OHS enforcement officers are not Carabinieri, which are known for their strict deterrence-driven approach to law enforcement (Nelken 2010). However, both OHS officers and Carabinieri have UPG powers and should, thus, behave in the same way when enforcing the law. This is exactly what the UPG law and CPP are for, to ensure legal consistency and proportionality, and the equality-before-the-law principle envisaged in the Italian Constitution, and according to the separation of powers doctrine.

A further issue that emerged was that SPSAL and ITL have different viewpoints in terms of using discretion during enforcement activities. While SPSAL enforcement officers, who are part of the Ministry of Health, mentioned that the legal obligation of reporting OHS breaches hindered them from conducting their work, the ITL officers, who are directed by the Italian Ministry of Work and Social Policies, attributed much more importance to the function of their legally binding UPG duties, which prohibit the use of discretion. Hence, an SPSAL enforcement officer argued that

> The firm might be missing a document, but if a ventilation system or a machine guard is missing it is different. I try to operate in a way that if a health and safety document is missing it remains a small shortcoming, unless the issue can easily lead to worse problems. (Italian interview 5, Line 380)

This type of comment was mentioned by four more SPSAL enforcement officers and represents a breach of UPG enforcement duties. This approach, however, would probably be very much appreciated by duty holders. When enforcement officers start judging an employer's regulatory violations, they violate UPG duties and breach the CPP, fail to promote the equality before principle and undermine the separation of powers doctrine. Another SPSAL enforcement officer argued that

> It's an obligation [to follow UPG procedures]. This is why, let's say, maybe it can become a limitation when we would like to suggest solutions, to use a preventive [compliance-driven] approach, which goes beyond the UPG functions. This can be limiting for us. (Italian interview 3, Line 73)

These two statements show that the Italian enforcement officers working for SPSAL are not happy with using a strict discretion-less, deterrence-driven approach during enforcement, and they suggest that when possible preventive compliance-driven approaches should be used instead. However, the language used by ITL officers, who are often escorted by Carabinieri police officers during their enforcement activities, is much stricter.

> The majority of the health and safety regulatory violations in work environments represent penal breach. The enforcement officer's function is to restore a juridical order of the irregular situation encountered. The verification of regulatory requirements that can result in penal breaches means that enforcement officers must be Judiciary Police Officials [UPG]. (Italian interview 10, Line 27)

The enforcement approach differences between these two institutions show that officers working in these organisations have very similar official mandates, but different views on their functions. In other words, the activities conducted by SPSAL officers tend to be framed by their original preventive institutional enforcement mandate, which should be achieved with compliance-driven policies. This might be interpreted as a form of regulatory capture, which leads SPSAL officers to not consider employers breaching OHS regulations as criminal in the same way as ITLs and Carabinieri do.

2   Summary

Conclusions can be drawn about the relationship between the use of discretion during OHS law enforcement activities and principles of legal consistency, proportionality and fairness. This is an important subject because the achievement of modern criminal justice system values, such as equality before the law and the separation of powers doctrine, is based on the consistency and proportionality of a state's law enforcement agencies and officers. The use of discretion during regulatory enforcement activities is a highly debated subject because, on the one hand, every interaction requires a level of discretionary decision, but on the other hand, too much discretion can cause injustice and the under- or over-criminalisation of certain crimes.

The ideas of ensuring enforcement consistency, proportionality and fairness were considered a serious matter by the enforcement officers in Britain and Italy, but consistency was defined differently between the two jurisdictions. British officers argued that the aim of these principles was equality before the law, but mainly based on the broad achievement of similar regulatory *outcomes* rather than similar *actions*. Because of this ambiguity in the way the principle of legal consistency is applied, the HSE has often been accused of being influenced and even captured by duty holders' interests. In this research project it has emerged that this might be the case, which might be a major cause of the under-criminalisation of OHS crimes. Italian officers, due to their obligation to follow UPG law and CPP in their professional law enforcement activities, were not as concerned as their British equivalents with regard to ensuring consistent and proportionate enforcement actions, and there was almost no reference to the fair treatment of duty holders. That is because the Italian enforcement policies and regulations (UPG and CPP) ensure that these principles are respected during their field operation, which creates fairness. This includes the law enforcement agents' prohibition of using discretion during inspections. Despite this, Italian interviewees mentioned that they enjoyed the discretion that they were allowed to use until 1994. Thus, in the HSE Enforcement Policy Statement, 'consistency' is officially intended in terms of equality before the law, but in reality it is meant in terms of achieving uniform regulatory outcome (see also Hawkins 2002). However, from these interviews it emerged that achieving the same regulatory outcomes seems to have priority over that of ensuring consistent enforcement responses across firms and between different types of crimes, which is the reason why Carson (1970a, b, 1979) argued that OHS crimes are under-criminalised.

British interviewees argued that enforcing the law through credible dialogues with employers is a technique that improves the enforcement institution's credibility as well as consistency and fairness. British enforcement officers mentioned that not doing so might weaken their authority among the regulated community. Therefore, discretion is an essential tool for maintaining the regulatory enforcement institutions' authority, which confirms Hawkins' (2002) findings. However, from the Italian interviews it emerged that instituting a credible relationship with employers is less significant because the idea of enforcing the law in Italy does not rely significantly on officers' opinions. In other words, Italian officers argued that employers do not need to argue with them about the enforcement decisions to be taken. Duty holders are expected to provide the information requested, and a lack of cooperation can be considered as an obstruction to justice. Hence, engaging in discussions is unnecessary, as this can also weaken the officers' authority and legitimacy as law enforcement institution. This is because enforcement policies and procedures are enacted by parliament, not by the enforcement institution, and these cannot accommodate or be tailored to duty holder's circumstances. This was further confirmed by Italian interviewees, who said that being prohibited from using discretion since 1994 had strengthened their authority, which is the opposite of what their British colleagues believed would happen. The contribution that this comparative analysis has brought to the literature is that the credibility of regulations and enforcement institutions does not only depend on whether the enforcement officers listen and cooperate with the duty holders as Hutter (1988) argues, but also on whether the enforcement institution is equipped with the appropriate powers, resources and authority to achieve their regulatory duty. This authority and institutional legitimacy is achieved through consistent enforcement decisions, not only within the scope of the OHS regulation, but across the whole spectrum of the criminal law.

Another aspect that emerged from the interviews is that the use of discretion in Britain might lead to a more lenient approach to OHS enforcement than in Italy. In the comparative analysis of these two jurisdictions, it emerged that the more discretion enforcement institutions are allowed to use, the more lenient the enforcement policies will become in the long term. This research study helps to highlight the striking difference between the under-criminalisation of powerful people and organisations and the over-criminalisation of the powerless (Reiman and Leighton 2010). Italian OHS enforcement institutions have no discretion to decide whether to prosecute firms or not, while British do. This means that before undertaking a prosecution in Britain, the enforcement institution can base their decision on aspects that are not legalistic in nature, such as the public interest test, the resources available and the likelihood of winning the case. The reasons for giving these powers to the enforcement institution, which is something that was mentioned by the British and Italian officers and in the literature, is because resources can be used more efficiently, and the enforcement institutions actions be more effective. This is a very efficient strategy, but not consistent. A prosecution ending with an acquittal is not only a waste of resources but also a failure in regulatory enforcement terms because duty holders will have no obligation to reform their conducts (Hawkins 2002). The HSE's commitment to achieve the highest number of guilty sentences demonstrates that their aim is not only to do justice, but to do justice inexpensively, which are calculations that Italian law enforcement institutions and public prosecutors within the PM cannot do.

There seems to be confusion over the meaning of efficient enforcement activities and legally efficient ones. Packer (1968) argues that legal *efficiency* is meant as the criminal justice system's effective ability to convict guilty defendants and acquit innocent ones. Packer (1968) does not mean that the financial efficiency of the system should be taken into account. Therefore, it is questionable whether the decision to prosecute should take into account resources and whether this should be taken by the enforcement institution at all. It is true that discretion and independent enforcement decisions have been a tradition in Britain, but it is unclear why the CPS has not been given the responsibility to take over OHS prosecution cases from the HSE and LAs. Yet even if the CPS was assessing OHS prosecutable cases, the same problem might endure given the absence in Britain of an equivalent of the Italian CP and CPP or the principle of legal obligation.[4] Given the due-process

---

[4]The legal principle, to remind the reader, forbidding prosecutors (or Magistrates) to interrupt a court prosecution if there is sufficient evidence of criminal conduct.

model traditionally embedded in the Italian legal system, the Italian interviewees disagreed with the British practices used to achieve both enforcement and financial efficiency in their operation. That is because constitutional principles prohibit the executive, which control the enforcement institutions, from taking decisions that should be taken by the judiciary. Italian enforcement officers argued that the HSE should not be given the responsibility to judge whether to prosecute incompliant firms. However, they also complained about the lengthy, and often fruitless, trials and the damaging effect that these issues have for regulatory compliance, but this is caused by inadequate funding of the enforcement institution and judiciary. The discretion embedded in the British enforcement policies together with under-funded enforcement institutions, in other words, causes the decriminalisation of OHS crimes. As a consequence, citizens' civil rights are not upheld, and the harm caused by OHS crimes is not prevented, punished or compensated adequately.

If regulation is defined as a means to repair collectively damaging externalities caused by free-market activities creating 'tensions between public aims and private interests' (Ogus 1994, p. 7), it can be argued that OHS regulation is failing. This is because at the moment the damaging externalities are borne mostly by workers and their families rather than businesses (HSE 2013). The redistribution of wealth from businesses to workers induced by the regulation decreases if employers are not fully legally liable for workers' OHS, or if workers are not compensated for incidents, but also if enforcement institutions are allowed the discretion to adapt their enforcement activities to specific situations and the resources available. This is also demonstrated by the HSE's priority to ensure compensation to victims and their families, rather than achieving a suitable conviction. The fierce attack waged on the OHS compensation culture pursued by the British Government in early 2010s[5] has further decreased victims' compensation rights, which means that British workers might be even less likely to be compensated for incidents (*Hazards Magazine* 2013). Due to the National Institute

---

[5]Health and safety compensations have decreased since 2000 (*Hazards Magazine* 2013).

for Insurance against Working Accidents (INAIL), the publicly funded OHS insurance administration, in Italy the situation seems better, but there is no statistics available to confirm this.

These findings also demonstrate that ambiguous economic regulations can also cause higher regulatory failures when used in conjunction with discretionary practices. Firstly, by not regulating business governance, which allows organisations, large ones in particular, to set up complex internal governing structures throughout the organisation to reduce directors' criminal liabilities. Secondly, by implementing vague and interpretative regulations, which increase the interpretability of rules, prevents enforcement institutions from employing easy and straightforward enforcement practices, and seriously challenges criminal prosecutors' ability to achieve guilty sentences during trials. Thirdly, by allowing enforcement institutions to cooperate with businesses and, by doing so, ignoring the conflict of interests this creates between duty holders and law enforcers. This conflict of interest might be exacerbated during political and economic crises, when the need to avoid firms' bankruptcy becomes more important than protecting workers' OHS. Hence, regulatory discretionary practices do not act on their own but interact with legal, political and social trends and lead to specific enforcement outcomes.

It can be concluded, thus, that discretion is power. Discretion empowers enforcement officers with the freedom to make subjective decisions. It empowers enforcement institutions to decide their enforcement actions and, to various degree, policies. It empowers the state's executive to take decisions that are unchecked by the legislative. Anyone using discretion is less regulated and, thus, more empowered, but the more discretionary powers are given to criminal justice system institutions, the more capture and inconsistency and unproportionate reactions to crimes it causes. This leads to social inequality and injustice. Hence, it is not better to allow and manage discretion, but to manage an effective way to reduce the level of discretion. The compliance-driven policies theorised and adopted in Britain in the past decades sems to have ignored this issue. This issue decreases the overall legitimacy of the criminal justice system. The interpretivist, labelling, pluralist or dramaturgical theoretical frameworks used by academics endorsing compliance-driven regulatory enforcement policies are myopic of the underpinning hierarchical social powers dictating the relationships between the state and citizens. In other words, at the moment law enforcement discretionary practices are used to support a capitalist economic system that disregards public interests for the sake of private capital accumulation.

References

Baldwin, R., & Hawkins, K. (1984). Discretionary justice: Davis reconsidered. *Public Law, 580*(Winter), 570–599.

Baldwin, R., & Veljanovski, C. G. (1984). Regulation by cost-benefit analysis. *Public Administration, 62*(Spring), 51–69.

Bardach, E., & Kagan, R. A. (1982). *Going by the book: The problem of regulatory unreasonableness*. Philadelphia: Temple University Press.

Bartrip, P. W. J., & Fenn, P. (1980a). The conventionalization of factory crime a re-assessment. *International Journal of the Sociology of Law, 8,* 175–186.

Bartrip, P. W. J., & Fenn, P. (1980b). The administration of safety: The enforcement policy of the early factory inspectorate 1844–1864. *Public Administration, 58*(Spring), 87–102.

Bartrip, P. W. J., & Fenn, P. (1983). The evolution of regulatory style in the nineteenth century British factory inspectorate. *Journal of Law and Society, 10,* 201–222.

Becker, H. S. (1963). *Outsiders: Studies in the sociology of deviance*. New York: Free Press.

Becker, H. S. (Ed.). (1964). *The other side: Perspectives on deviance*. New York: Free Press.

Black, J. (2001). *Managing discretion*. Paper presented at the ARLC conference [Online]. Available from: http://www.lse.ac.uk/collections/law/staff%20 publications%20full%20text/black/alrc%20managing%20discretion.pdf. Accessed 25 Aug 2014.

Braithwaite, J. (1985). *To punish or persuade: Enforcement of coal mine safety*. Albany: State University of New York Press.

Braithwaite, J. (1987). Negotiation versus litigation: Industry regulation in Great Britain and the United States. *American Bar Foundation Research Journal, 2,* 559–574.

Braithwaite, J. (1997). On speaking softly and carrying big sticks: Neglected dimensions of a republican separation of powers. *University of Toronto Law Journal, 47*(3), 305–361.

Carson, W. G. (1970a). White collar crime and the enforcement of factory legislation. *British Journal of Criminology, 10*(4), 383–398.

Carson, W. G. (1970b). Some sociological aspects of strict liability and the enforcement of factory legislation. *Modern Law Review, 33,* 396.

Carson, W. G. (1979). The conventionalisation of early factory crime. *International Journal of the Sociology of Law, 7,* 37–60.

Centre for Corporate Accountability (CCA). (2008). *Fines against most companies convicted following work-related deaths less than 1/700th of their turnover, new research shows* [Online]. Available from: http://www.corporateaccountability.org.uk/press_releases/2008/mar16sent.htm. Accessed 25 Aug 2014.

Cicourel, A. V. (1968). *The social organization of juvenile justice.* New York: Wiley.

Damaška, M. R. (1973). Evidentiary barriers to conviction and two models of criminal procedure: A comparative study. *University of Pennsylvania Law Review, 506,* 1972–1973.

Davis, C. (2004). *Making companies safe: What works?* Centre for Corporate Accountability [Online]. Available from: http://www.unitetheunion.org/uploaded/documents/Making%20Companies%20Safe%20-%20what%20Works%20(CCA-Unite%20paper)11-4856.pdf. Accessed 25 Aug 2014.

de Secondat, C.-L., & Baron de La Brède et de Montesquieu. (2001). *The spirit of the laws.* (Translated from the French, by D. W. Carrithers & T. Nugent) Kitchener, ON: Batoche Books (Originally printed in 1748).

Dworkin, R. (1977). *Taking rights seriously.* Cambridge: Duckworth Press.

Emerson, R. M. (1969). *Judging delinquents: Context and process in juvenile court.* Chicago: Aldine.

English Oxford Dictionary. (2017). Discretion. In *The English Oxford dictionary* [Online]. Available from: https://en.oxforddictionaries.com/definition/discretion. Accessed 17 Aug 2017.

Fioravanti, M. (2011). Le dottrine dello stato e della costituzione. In R. Romanelli (Ed.), *Storia dello Stato Italiano dall'unità ad Oggi.* Roma: Donzelli.

Fooks, G. (2008). *The relationship between the levels of fines imposed upon companies convicted of health and safety offences resulting from deaths, and the turnover and gross profits of these companies.* Centre for Corporate Accountability [Online]. Available from: http://www.corporateaccountability.org.uk/dl/manslaughter/reform/ccasentresearchmar08.doc. Accessed 25 Aug 2014.

Galligan, D. J. (1986). *Discretionary powers.* Oxford: Oxford University Press.

Goffman, E. (1959). *The presentation of self in everyday life.* Garden City: Doubleday Anchor.

Goffman, E. (1961). *Encounters: Two studies in the sociology of interaction.* Indianapolis: Bobbs-Merrill.

Goffman, E. (1963). *Behavior in public places: Notes on the social organization of gatherings.* New York: Free Press.

Goffman, E. (1967). *Interaction ritual: Essays on face-to-face behaviour.* New York: Anchor Books.

Goffman, E. (1970). *Strategic interaction.* Oxford: Blackwell.

Goffman, E. (1971). *Relations in public: Microstudies of the public order.* Harmondsworth: Penguin.

Gunningham, N., & Johnstone, R. (1999). *Regulating workplace safety: System and sanctions.* Oxford: Oxford University Press.

Gustapane, A. (2012). *Il ruolo del pubblico ministero nella Costituzione italiana.* Bologna: Bononia University Press.

Hawkins, K. (1984). *Environment and enforcement: Regulation and the social definition of pollution.* Oxford: Clarendon Press.

Hawkins, K. (2002). *Law as last resort: Prosecution decision-making in a regulatory agency.* Oxford: Oxford University Press.

Hazards Magazine. (2013, April–June). Robbed! Bloody bandages but no bloody compensation. *Hazards Magazine,* Issue 122 [Online]. Available from: http://www.hazards.org/votetodie/robbed.htm. Accessed 25 Aug 2014.

HSE. (2013). *Health and safety statistics* [Online]. Available from: http://www.hse.gov.uk/statistics/index.htm. Accessed 25 Aug 2014.

HSE. (2015). HSE *enforcement policy statement* [Online]. Available from: http://www.hse.gov.uk/pubns/hse41.pdf. Accessed 31 Jan 2018.

Hutter, B. (1988). *The reasonable arm of the law? The law enforcement procedures of environmental health officers.* Oxford: Clarendon Press.

Hutter, B. (1997). *Compliance: Regulation and environment.* Oxford: Clarendon Press.

Justice Committee. (2009). *The crown prosecution service: Gatekeeper of the criminal justice system* (Session 2008–2009, 9th Report) [Online]. London: Justice Committee Publications. Available from: http://www.publications.parliament.uk/pa/cm200809/cmselect/cmjust/186/18602.htm. Accessed 25 Aug 2014.

Langbein, J. H., Lerner, R. L., & Smith, B. P. (2009). *History of the common law: The development of Anglo-American legal institutions.* New York: Aspen Publishers. ISBN 978-0-7355-6290-5.

Legrand, P. (1996). European Legal Systems Are Not Converging. *The International and Comparative Law Quarterly, 45*(1), 52–81.

Manning, P. K. (1977). *Police work: The social organization of policing.* Cambridge: MIT Press.

Meindinger, E. (1986). Regulatory culture: A theoretical outline. *Law and Policy, 9,* 355–386.

Mousourakis, G. (2015). *Roman Law and the Origins of the Civil Law Tradition.* New York, Dordrecht and London: Springer International Publishing. ISBN 978-3-319-12267-0; e-ISBN 978-3-319-12268-7; https://doi.org/10.1007/978-3-319-12268-7.

Nelken, D. (2010). *Comparative criminal justice: Making sense of difference.* London (UK), Thousand Oaks (CA), New Delhi (IND) and Singapore: SAGE Publications Ltd.

Newburn, T., & Reiner, R. (2007). Policing and the police. In M. Maguire, R. Morgan, & R. Reiner (Eds.), *The Oxford handbook of criminology* (4th ed.). Oxford: Oxford University Press.

Ogus, A. (1994). *Regulation: Legal form and economic theory.* Oxford: Clarendon Press.

Olsen, P. (1992). *Six cultures of regulation.* Copenhagen: Handelshojskolen.

Packer, H. L. (1968). *The limits of criminal sanction.* Stanford, CA: Stanford University Press.

Pearce, D. W. (1983). *Cost-benefit analysis.* London: Macmillan.

Posner, E. (2003, December). Transfer regulations and cost-effectiveness analysis. *Duke Law Journal, 53*(3), 1067–1079.

Reiman, J., & Leighton, P. (2010). *The rich get richer and the poor get prison: Ideology class and criminal justice.* Boston: Allyn & Bacon.

Reiss, A. J. (1971). *The police and the public.* New Haven, CT: Yale University Press.

Richardson, B., Ogus, A., & Burrows, P. (1983). *Policing pollution: A study of regulation and enforcement.* Oxford: Clarendon Press.

Ross, H. L. (1970). *Settled out of court: The social process of insurance claims adjustment.* Chicago: Aldine.

Sabine, G. H., & Thorson, T. L. (1973). *A history of political theory.* Hinsdale: Harcourt Brace.

Skolnick, J. H. (1966). *Justice without trial: Law enforcement in democratic society.* New York: Wiley.

Sudnow, D. (1965). Normal crimes: Sociological features of the penal code in a public defender office. *Social Problems, 12,* 255–276.

Tombs, S., & Whyte, D. (2007). *Safety crimes.* Cullompton: Willan Publishing.

Tombs, S., & Whyte, D. (2009). A deadly consensus: Worker safety and regulatory degradation under New Labour. *British Journal of Criminology, 52*(5), 997–1016.

Tombs, S., & Whyte, D. (2010). *Regulatory surrender: Death, injury and the non-enforcement of law.* Liverpool: Institute of Employment Rights.

Vile, M. J. C. (1963). *Constitutionalism and Separation of Powers.* Oxford: Clarendon Press.

7

Discretion and Enforcement Activities

This chapter analyses, compares and discusses how law enforcement discretionary practices affect the enforcement institutions and officers during various stages of their jobs. It analyses how discretion affects the work of enforcement officers in practice, and how this can lead to the under-criminalisation of occupational health and safety (OHS) crimes. Chapter 6 and this chapter come in tandem. While Chapter 6 analysed and compared how the use of discretion has severe implications for the achievement of the modern criminal justice system's values, this chapter aims to analyse the technical issues of allowing the use of discretion by OHS enforcement officers and regulators.

This chapter is divided into three sections where both the British and Italian responses are analysed and compared together. The first section analyses and compares how discretion affects the proactive enforcement activities planning process in both jurisdictions. This is important because enforcement institutions' proactive targeting decisions affect the efficacy of their work and, thus, the criminalisation of duty holders. The second section analyses the issues faced by enforcement officers when they are empowered to enforce the law with both deterrence-driven policies and compliance-driven ones. The compliance-driven enforcement

© The Author(s) 2018   207
D. Canciani, *The Politics and Practice of Occupational Health and Safety Law Enforcement*, Critical Criminological Perspectives,
https://doi.org/10.1007/978-3-319-98509-1_7

traditions of the institutions in both jurisdictions create conflicts between the officers' willingness to achieve compliance and the requirements of the legal system to achieve justice. Hence, the last section analyses how enforcement officers perceive on-the-spot penalties and enforcement-related charges in both jurisdictions.

1   Discretion and Planning Enforcement Activities

A common issue that emerged during both the Italian and British interviews was how having the discretion to plan proactive enforcement activities and targets can affect workload and the level of criminalisation of OHS crimes. The number and firms visited depend on the amount of discretion they are allowed to use. In both countries local office directorates are assigned a number of inspections to undertake per industry classification on a yearly basis, which are also influenced by the risks of economic sectors, national priorities, resources and local territorial characteristics. During interviews, officers argued that planning proactive enforcement activities is not a straightforward process. On the one hand, targeting strategies should concentrate on the riskier and most harmful sectors of the economy, while on the other they should also target all duty holders equally and consistently. The main reasons for modifying the enforcement targeting imposed nationally by the chief officers were numerous, but the main reason was related to specific local territorial economic characteristics and the need to achieve the annual enforcement targets. Therefore, although the use of a targeting strategy is used to ensure a standard and proportionate enforcement action across geographical areas, both Italian and British enforcement officers complained that the targets imposed are not flexible enough, are not monitored enough, and led to a significant lack of enforcement activities in certain geographical areas. Again, in this instance, empowering officers with more discretion to plan enforcement targets could improve enforcement outcomes, but allowing too much discretion could lead to inconsistency.

In both countries, enforcement officers complained that the top-down enforcement targets and priorities might represent an issue when planning the enforcement activities. Hence, this is what an Italian enforcement officer argued while talking about the 5% of firms that they are required to visit annually.

What sense does it make to give an industrial definition [of the enforcement targets] at national level, when at local level the situation might be completely different? So, let's keep the targets rigid, but specify that if the local circumstances are different, you change the targeting specifically to your local needs. [...] If a local economy is comprised mostly of hotels, such as in the Marche region or on the coast, would you impose an 8% target on woodwork carpenters? You should require instead enforcing hotels' structural risks and manual handling. (Italian interview 11, Line 236)

This officer agrees that setting national enforcement targets ensures consistency, but also that these targets can be meaningless in practice. They stated that more discretionary decision powers might make their work at local level better because it would improve targeting techniques and become more effective in ensuring compliance and preventing OHS crimes. However, giving responsibility to chief enforcement officers to choose the firms to target might lead to irrational or ineffective enforcement. Again, balancing the amount of discretion enforcement officers would like to have and the amount they should have represented a conflict of interests.

The planning of enforcement activities in Britain occurs through a bidding system between the Health and Safety Executive (HSE), Field Operations Directorate (FOD) sectors and the Cross-Cutting Intervention Department, which jointly agree on the annual enforcement targets for each HSE operational sector. However, when it comes to selecting specific firms, a British enforcement officer argues that:

The previous administration was probably content with the idea that had grown up historically, of visits on a sort of time basis roughly related to the perceived risk of the previous inspector. You know, this company was category A, B or C. A receives more visit and more frequently than C did. This may change. This government does not want us to work on such as time bases. (British interview 2, Line 363)

Local enforcement offices are required to select firms based on an internal qualification of their risk of breaching OHS regulations. The British targeting process seems more structured than the Italian one, at least until 2011 when the last interviews were conducted. They are required to respond for the firms selected and justify rationally why they targeted specific firms, which is not required from their Italian colleagues. It is worth a reminder from Chapter 6 that a major difference between the two jurisdictions is that in Britain the total number of annual enforcement inspections can change, also according to the institution funding, while in Italy enforcement policies cannot change and the annual enforcement overall target is 5% of firms in the territory. Hence, the British enforcement institutions and officers have decisional flexibility, which would be welcomed by their Italian counterparts. However, Italian chief enforcement officers' discretion to choose organisations to target during field operations might be given to compensate for their lack of discretion to choose enforcement policies and overall annual enforcement targets. The same British officer argued that he has

> come to some decisions about how much and what I can contribute to [in terms of enforcement targets], and there will be things that I know I have to contribute to. Work that we were doing this year on LPG [Liquefied Petroleum Gas], because of the explosion in Glasgow a few years ago, has been the top priority for our proactive work. And in some parts of the country that has been pretty much all lived on proactively because there were so many sites to visit. But in others it was less of a requirement. In some areas there are not as many LPG visits, but other Field Operator Directorates are more discretionary. So they have taken forward other local priorities. (British interview 2, Line 112)

This interview was conducted at the start of 2011, but with the launching of the 2011 Lord Young's *Good health and safety, good for everyone* strategy, the targeting technique might have changed to account for the 35% reduction of funding and a strong commitment to reduce proactive activities. British OHS enforcement institutions planning and targeting policies allow the HSE to reduce these activities as much as they have to cope with less funding, which might further criminalise these crimes, but this is not the case in Italy because enforcement policies and overall annual targeting must be agreed in Parliament and comply with constitutional values.

In Italy, OHS chief enforcement officers are not allowed to choose how much to enforce, or enforcement policies, but they can use more discretion to choose the firms to visit. An Italian chief enforcement officer argued that

> Enforcement activities are based on the achievement of enforcement targets […] This depends on the enforcement officer, because unfortunately […] we are forced to visit a determined number of construction sites. But no one tells us to go to specific building sites where there is a specific risk… We could visit three hundred construction sites per year without issuing one penalty because they are found compliant. But we know that they are compliant before visiting them. Or we could go to three hundred construction sites that we know will not be compliant. That happens because we can select the construction sites. (Italian interview 13, Line 376)

This targeting technique used by the Italian enforcement institution gives officers discretion to decide which firms to visit, which might not be based on the level of risk of the firm but just used to meet annual targets without contributing to improving overall regulatory compliance. This flexibility might also allow officers to manage workload when, for example, reactive investigations increase. This was another common complaint among enforcement officers. If you just need to meet targets, and this system pushes the enforcement officers to visit complying firms to achieve that, then the enforcement activities might not be effective enough to improve working conditions. Hence, Italian chief enforcement officers are entrusted with the unchecked power to choose the firms to visit. This is not the case in Britain, where a risk-based approach is used across the different directorates.

Therefore, allowing more enforcement targeting discretion is ideal for managing enforcement workloads, but this confers on chief enforcement officers greater responsibility and trust to pick the right firms and improve the effectiveness of their work. During the interviews this commitment was shown extensively among the enforcement officers of both countries. However, the amount of discretion allowed while targeting firms represents a continually negotiated balance, which is difficult to measure, and the results are difficult to account for. The system used to select enforcement targets in Britain deprives chief enforcement officers of their discretionary powers, and increases their accountability, while the targeting system in Italy gives great discretion to chief enforcement officers, who are not required to report and, hence, be overly accountable, for their field operations.

Conclusions can also be drawn on the relationship between the use of discretion and the planning of proactive enforcement activities conducted by OHS enforcement officers. While in Britain these decisions are undertaken jointly by the HSE and the various local FODs, in Italy these decisions are guided by national targets, but are ultimately left to the local chief enforcement officers to decide. In fact, from the Italian interviews it emerged that enforcement activity workloads can be greatly affected by these decisions. Local chief enforcement officers might choose firms known to be compliant in order to achieve annual targets. The issue is that the work of the enforcement officers might not concentrate on the most dangerous sectors of the economy and thus be ineffective to improve compliance or detect crimes. From the interviews it emerged that Italian OHS enforcement officers must be trusted with a professional commitment to improve working conditions, which is essential to ensure a higher quality of the enforcement activities conducted. Therefore, using a system like the British one is better, because annual enforcement targets are decided through the HSE headquarters and local enforcement offices' agreements. This also allows for tailoring enforcement activities to specific geographical economic needs.

2   Discretion and the Use of Compliance-Driven Enforcement Policies

The amount of technical information and guidance officers can give to employers during enforcement, which aim to achieve more compliance and prevent incidents, was also a topic that emerged during the research study in Britain and Italy. Discretion and the level of separation of powers between state institutions play an essential role while trying to explain the two approaches. That is because the enforcement institutions, if they have the power to negotiate the policies to use, such is the case in Britain, can decide whether OHS breaches should be enforced with compliance- or deterrence-driven policies. If enforcement policies are not regulated by strict enforcement procedures, the institution and officers will be able to adapt their behaviours to different contexts more easily.

Compliance-driven policies, in the current socio-economic global context, can aid duty holders in achieving a competitive economic advantage and avoid conflictual and expensive enforcement relations with the state. Without a strict enforcement regulation and discretion, enforcement institutions can easily start adopting compliance-driven policies over deterrence-driven ones, but this creates social injustice because the state response to crimes discriminates certain social classes over others. If enforcement policies are more strictly regulated according to non-discretionary law enforcement procedures, which can include the use of compulsory notices and on-the-spot penalties such is the case in Italy, compliance-driven policies are less likely to be used during proactive enforcement activities because otherwise officers' behaviour would be in breach of law enforcement regulation. In Britain, this is less the case. Given that in Britain the use of discretion is allowed and in Italy it is not, OHS enforcement officers provided interesting responses when questioned about the use of compliance-driven policies during proactive enforcement activities.

The question is how much information or help would an enforcement officer give to duty holders, before or after issuing an improvement or prohibition notice? And to what extent would regulatory failure lead to a charge for the enforcement officer's intervention (i.e. an), or to start a prosecution? While explaining that Italian inspectors are strictly forbidden from working as consultants, a British enforcement officer argued the following.

Yes, in principle we have the same rule. What we say to our inspectors is that you're not there to be a free consultant, but at the same time, when we see something wrong, we would expect our inspector to tell people broadly how to put it right. So, in terms of a machine that needs a guard, we would expect the inspectors to describe roughly the sort of guard, the size of the guard, the type of material and interlocks. We would not expect them to actually tell them in detail how to actually wire the electric interlock switch. They have their own expertise for that, but we would expect the inspectors to be able to say that you want a guard that comes down that far, with a gap of no more than X. So, yes we are not there to be free consultant for them, but we expect them to give the employer enough information so that they know what they are supposed to do. Sometimes it's a very fine balance. (British interview 3, Line 953)

While talking about the difficulties of enforcing regulations, another British officer explained that

[We] wear two hats. That is one of the major problems with the job. We are there to inspect and advise and help people, but if we are investigating an accident we are looking to prosecute, especially if it is a fatal accident. (British interview 1, Line 282)

These enforcement officers are identifying the issue. Enforcement officers' responsibility is to enforce the law, but deciding when to use compliance- or deterrence-driven policies is difficult. Officers also have valuable professional experience that can help employers understand how to reach regulatory compliance. Officers should not provide information, but it would be absurd to issue a notice or a penalty and leave the duty holders' premises without providing any basic guidance on how to reach regulatory compliance. Hence, there is a case for enforcement officers to play the role of advisers, but the decision of providing information to employers seems to be a practice that requires a level of discretionary judgement from each enforcement officer. The introduction of the Fee for Intervention (FFI) policy in Britain has partially resolved the issue. Duty holders receive information from the enforcement officers for a fee. What is unknown, however, is how much regulatory breach corresponds to a *material* breach of the law, whether that is applied consistently across the territory, and how much information enforcement officers are expected to give to duty holders.

Italian enforcement officers are not allowed to give information to employers because it creates conflicts of interest and unfair economic advantages to those firms that can afford to pay over those that cannot. Hence, the Italian Code of Penal Procedure (CPP) and Judiciary Police Official (UPG) laws forbid charging fees and forces enforcement officers to issue on-the-spot penalties when breaches are detected. They argued that there is a very narrow difference between working as law enforcement officers and as advisors and trainers. Enforcement officers are not allowed to be consultants at any time during their enforcement duties, but they can when they are not Judiciary Police Officials (UPGs). Training and enforcement are two separate aspects and distinct moments of enforcement officers' job duties. They can be technicians and advisors when organising conferences and seminars addressed to stakeholders, but not when conducting enforcement activities as UPG. Hence,

> Assistance for example, means to increment all of the safety processes in the firm, which also involves SPSAL (Prevention Services for Safety in Work Environments). In that case, Judiciary Police (UPG) functions are not used. But if I use the Judiciary Police function hat when entering a firm, I also do it as a means to achieve prevention. Hence, we avoid the occurrence of incidents by repressing violations. (Italian interview 5, Line 35)

It is noteworthy that in both countries enforcement officers used the analogy of wearing two types of enforcement hats, which is used to explain two distinct aspects of enforcement activities. However, while the British officers' roles might change during the course of an inspection, Italian officers' roles cannot.

A Labour Territorial Inspectorates (ITL) enforcement officer also mentioned this dual function when arguing that

> The priority is prevention. Instructing and surveillance are two different actions when ensuring prevention [as a regulatory system]. Consultancy is forbidden. Information dissemination and training situations occur through special gatherings aimed at workers and employers. (Italian interview 10, Line 52)

ITL enforcement officers seemed to be more conscious of the importance of keeping deterrence- and compliance-driven enforcement activities separate, but, as seen earlier, they also significantly valorise the provision of information and education. The preventive goal of the Italian national health service is also embedded into the reasoning of SPSALs officers. Hence, Italian interviewees were also asked the difference between an advisory role and a consultancy role. A SPSAL officer argued

[We] risk becoming consultants, which is totally prohibited in our legislation. Because if I intervene in something generic, it's all right, I can give information, I can assist. But if we start resolving specific problems, we become consultants and this is prohibited. So, there is this willingness from all of us to not only be police officers, but to be technicians, and this is where the problems begin. We do it, but we always do it, perhaps, after a violation has been detected and reported [to the public prosecutor]. (Italian interview 3, Line 103)

From this quote, it becomes clear that the definition of consultancy and advice depends on the amount of information given. Advice consists of replies to specific questions, while consultancy is providing comprehensive solutions to complex health and safe work issues. In addition, in Italy, brief specific replies on how to enforce the regulation can only be given after employers have been issued with a penalty. During pre-organised seminars, conferences and in-office face-to-face meetings, the amount of information that can be given to duty holders is potentially unlimited. The decision to behave like police officers or helpful technicians depends on the gravity of the violation or the numbers of breaches found. Hence, the moment officers decide to stop being technicians and become officers depends on many different factors.

This issue also prompted an enforcement officer to suggest a system that might resolve this issue.

Ten metres away from us we could have a consulting service [...] But the question is always the same: is it possible to combine the consultant and the inspector inside the same institution? Maybe we should create this service under a different institution, hence end up with a consulting institution and an inspecting institution. (Italian interview 7, Line 183)

While ITL officers see deterrence-driven enforcement as a central aspect of their enforcement activities, SPSAL officers' institutional culture emphasise much more the use of compliance-driven enforcement practices. The officers' suggestion to separate deterrence-driven and compliance-driven policies between different institutions is suggested as an option in the 1989 European Economic Community (EEC)

Council Directive on health and safety (89/391/EEC).[1] The directive, however, does not specify whether deterrence- and compliance-driven practices should be performed by the same institution. The regulation remains ambiguous because different European Union (EU) jurisdictions regulate law enforcement institutions' powers and officers' actions differently. This mostly depends on the separation of powers doctrine envisaged by the state in each jurisdiction and expressed through, in this case, the HSE Enforcement Management Model (EMM) and Italian CPP. Hence, while in Britain the HSE has more flexibility to decide the enforcement approach, in Italy this is not the case, and since 2000 the National Institute for Insurance against Working Accidents (INAIL) has increasingly acquired the responsibility to promote regulatory compliance through, mainly, compliance-driven policies.

3   Discretion and On-the-Spot Penalties

The use of administrative on-the-spot penalties or charges by enforcement officers during inspections was another theme that emerged from the interviews with enforcement officers in both jurisdictions. Enforcement officers' powers to issues penalties in both jurisdictions converged. While, the Italian SPSAL were required to issue compulsory on-the-spot penalties from 1994, in October 2012 the HSE started to charge employers with fees (i.e. FFI) when detected with material breaches of the law. These are different tools because in Italy these are legal tools, while in Britain these are charges, but also similar because they are issued to employers when in breach of the regulation and because they achieve similar outcomes.

Enforcement officers in both jurisdictions identified the power to issue administrative penalties as a positive aspect. Firstly, they mentioned that it encourages duty holders to reach regulatory compliance *before* enforcement officers' inspections. Secondly, they stated that sanctions give an economic competitive disadvantage to non-complying

---

[1]Council of the European Union. (1989). *EEC Council Directive 89/391/EEC.* Introducing measures to encourage improvements in the health and safety of workers at work. Official Journal L 183, 29 July 1989, p. 1–8.

firms, and hence promotes a culture of regulatory compliance. Finally, sanctions might increase the efficacy and efficiency of the enforcement actions, because of the increased deterrence caused by the penalties and revenues collected. However, the idea of adopting administrative sanctions in Britain and Italy also attracted negative opinions.

An interesting observation that officers from both jurisdictions made was that the reason for not issuing penalties in Britain and issuing them in Italy was to comply with the state separation of powers doctrine, and while British officers thought charging firms is fine, Italians believed it would create conflict of interest, transform them into consultants and undermine their roles as law enforcers. In Britain, enforcement officers argued that they can only charge employer a fee for fixing a regulatory breach because these are not judiciary actions like penalties. They are prohibited from issuing penalties because the executive institutions should not judge legal breaches, and hence punish or penalise citizens without Judiciary Authority's approval. Officers argued that on-the-spot administrative penalties have been criticised because they might lead to over-criminalisation and because they might be used by the executive to raise revenues. Italian officers cannot charge employers fees because their actions would create a conflict of interests and contravene UPG law enforcement principles, but the law gives them the power to issue on-the-spot penalties, which is regulated by Parliament and supervised and approved by the PM. Hence, in both countries officers were aware of the separation of powers conflict that issuing on-the-spot penalties might lead to, but, unlike in Britain, the Italian legal system reduces this conflict by making officers directly answerable to the Judiciary Authority and, in particular to the PM. After all, charging firms for a service penalises them, but it also deprives the Judiciary of the control to supervise inspectors punitiveness during enforcement.

When talking about the possibility of introducing on-the-spot administrative sanctions, a British enforcement officer argued that

> There are problems with being able to issue on-the-spot fines. I think our strength as regulators is our independence. As soon as you start to issue fines and bring in revenue and generate income, then there are questions over our independence. So, whereas before we would issue an instruction because clearly there was a problem, as soon as you start to introduce

income revenue, the obvious question is going to be: are you taking that action because you think is necessary, or you are taking that action because it is going to bring in some money? (British interview 6, Line 280)

However, when prompted by the suggestion that the money collected by the enforcement officers might just go to the Treasury rather than to the inspectorate, the officer insisted that in 'the new consultation on fees-for-failure the money would not go to the Treasury, the money would go to the HSE' (British interview 6, Line 299). An immediate question this response induces is why should the HSE maintain control over the revenue made through the FFI policy? The FFI policy has been enforced since 2012, which allows the HSE to collect fees from stakeholders to pay for their intervention during inspections. To obviate the conflict of interests caused by the HSE direct collection of fees, the HM Revenue and Custom forbade the HSE to keep more than $11 million collected through the FFI policy. This is interesting because capping the fees collectable by the HSE reduces the institution's effectiveness of its enforcement activities, and hence, its institutional independence.

From the interviews it emerged that introducing FFI policy changes the relationship between enforcement officers and duty holders. Hence, while talking about this subject, a British enforcement officer argued that they had

spent ten years with the chemical industry, but before [the fees charges[2]] regime came into force, and I think that it has totally changed the dynamic between the duty holders and the inspectors. (British interview 3, Line 869)

When questioning whether the charging system might become a sort of consulting service, the interviewee disagreed, but argued that a fee would attract opposition from large organisations and fear from small ones. The officer also predicted that the FFI policy might have a significant impact on businesses as most employers are usually found in material breach of the regulation.

Given that enforcement officers complained that penalties are generally too low, the expectation is that the introduction of administrative

---

[2]Some industries in England were already paying an enforcement fee before 2012.

sanctions might also be supported. However, this did not seem to be the case among the British enforcement officers:

> You talk with some politicians in Britain and particularly in the current government, who are anti-health and safety, and their impression is that most people comply and there is an only small percentage of people who are not complying with the legislation, and quite frankly they do not know what they are talking about. Twice in my career I have been into a workplace where there was effectively nothing wrong. Twice in forty years. (British interview 3, Line 112)

This interviewee depicts quite clearly the frustration that enforcement officers can have towards the Government and the reforms that have affected the HSE in the past years. In addition, according to this participant the amount of non-compliance is quite extended and the charging system, which was introduced roughly the year after this interview took place, might become financially onerous for businesses and the economy.

It is worthwhile notiinig how a judicial sanctioning system, such as the one used in Italy, and discretion bear the same legal issues in terms of the separation of powers doctrine in Britain, but lead to two opposite responses from the enforcement officers. On the one hand, the FFI charging system was criticised because of fears that the HSE could lose its independence and because it might be perceived as a means to raise funds and, by doing so, damage the institution's reputation. Officers also argued that the Judiciary would disapprove judicial on-the-spot penalties because the executive, through the HSE, would acquire too much power and become a threat to civil liberties. On the other hand, enforcement discretion was appreciated because it preserves the independence of the institution. Yet, British officers did not see the HSE discretion as an issue, even though it is influenced by the government through the level of funding given, can have a significant impact on stakeholder's civil liberties, and on workers and victims. Discretion, hence, also has the potential to damage the institution's reputation. The reason discretion was not considered an issue is because until 2012 the HSE could not penalise duty holders directly. The HSE could penalise duty holders only after court trials, which means that the Judiciary would be involved in that decision. However, the system until 2012, under-criminalised OHS crimes. This is the main cause why the HSE has been accused of being subject of conflicts of interests and captured by duty holders' interests.

When looking critically at these responses, it can be argued that a penalising or charging duty holders for breaches is convenient for enforcement institutions. It is easier to raise revenue from sanctions and improve enforcement deterrence, than it is to risk resources through criminal prosecution against firms, such was the case in Britain until 2012. However, one of the major issues identified in Britain is that the enforcement institutions' income generated by the FFI charges creates conflicts of interests for the institution. Enforcement officers argued that charging companies with fees would not create a client relationship with duty holders, but will increase deterrence and improve the efficiency of their field operations. It has yet to be seen what will happen, but since 2012 the HSE FFI policy has gradually increased, which shows that there was an institutional willingness to recover from the recent funding cuts while improving the institution's regulatory deterrence capabilities.

Italian enforcement officers, as their British colleagues, appreciate using administrative penalties because it is a quick fix to non-compliance. However, the Italian enforcement officers could not see why issuing administrative sanctions might represent an issue for the separation of powers doctrine. When the Italian interviewees were told the motivations mentioned by the British officers, they argued that those issues do not apply in Italy. Firstly, the concern that the Judiciary Authority would lose control over law enforcement of OHS crimes in Italy does not apply. That is because public prosecutors' oversight the issuing of penalties of SPSAL's officers. This is also used to ensure that duty holders are punished and penalised consistently. Administrative sanctions might be used by governments to raise income, but this was not a concern in Italy because penalties levels are approved through acts of Parliament and income generated is going back to the Italian Treasury.

Italian officers complained that the level of penalties is an issue when trying to achieve compliance without discriminating between SMEs and larger firms. Italian officers argued that penalties are definitely too low for the legal breaches and creating deterrence, but only in larger organisations. They complained that small and medium-sized firms struggle to pay penalties and were worried these might bankrupt SMEs. Italian officers argued that the level of penalties should remain the same to avoid bankrupting SMEs, but be issued more frequently to create more deterrence. However, penalties can be issued more frequently only if the number of inspectors and inspections increase, which has not been the case for a long time.

Several interviewees argued that the introduction of on-the-spot penalties in 1994 significantly changed the way that the enforcement officers operated and the reactions of the duty holders. On one occasion an Italian enforcement officer argued that

> Until 1994, when Legislative Decree 626 came into force we were a team […] of multidisciplinary public consultants, who were very well trained and offered an inexpensive service. I mean, we were going into a factory, we were analysing it, and we were exposing unsafe working processes and machineries, and industrial hygiene issues. We were designing the improvements to apply and were delivering the reports to the duty holders and trade unions […] Then the Legislative Decree 626 1994 told us 'you are the enforcement institution'. […] After the Legislative Decree 626 1994, these aspects of, let's say, legal impact became routine and now constitute the core element of our work. So, to carry on working for the institution I had to brainwash and reprogram myself. (Italian interview 5, Lines 212, 215 and 358)

In this case, the discontinuation of the discretionary powers and the introduction of a more deterrence-driven approach to enforcement meant that the officers had lost part of their consulting and assisting powers and changed the nature of their duties, by starting to penalise duty holders with on-the-spot penalties. During the interviews the enforcement officers who experienced this significant change always showed melancholy about the old enforcement policies, because from 1994 onward, duty holders' perception of enforcement officers changed from being helpful consultants, and hence fully exercising the institutional preventive ethos, to punishers who could only partially help firms and workers improve working conditions while inspecting.

4 Summary and Discussion

Some key conclusions can be drawn on the relationship between the use of discretion and allowing enforcement officers to plan enforcement activities, choose between the use of compliance- and deterrence-driven policies, and allow them to use on-the-spot penalties or charge duty holders.

The first section of this chapter analysed and compared the relationship between the use of discretion and the planning of proactive enforcement activities conducted by OHS enforcement officers. While in Britain these decisions are undertaken jointly by the HSE and the various local FOD, in Italy these decisions are guided by national targets, but the local chief enforcement officers can decide the actual firms to inspect. In fact, from the Italian interviews it emerged that enforcement activity workloads can be significantly affected by these decisions. Local chief enforcement officers might choose firms known to be compliant in order to reduce workload and achieve annual targets. The issue here is that the work of the enforcement officers might not concentrate on the most hazardous sectors of the economy and thus be ineffective to improve compliance and detect breaches, which can thus under-criminalise OHS crimes. It emerged that Italian OHS enforcement officers must be trusted with a professional commitment to improving working conditions, which is essential to ensure a higher quality of the enforcement activities conducted. Therefore, using a system like the British one is better, because annual enforcement targets are decided by the headquarters and local enforcement offices' agreements. This also allows for tailoring enforcement activities to specific geographical economic needs or hazzards. Again, it can be concluded that less discretion has the potential to increase fairness and improve regulatory compliance levels.

From the comparative analysis it emerged that the institution's freedom to use discretion is a fundamental aspect when deciding the balance between deterrence- and compliance-driven policies. However, the extent and use of discretion is linked to the extent of the separation of state power between institutions. The European legal framework allows enforcement institutions to enforce the law with a mixture of compliance- and deterrence-driven policies, but only suggest that these two types of enforcement activities should be conducted by different institutions. In this comparative analysis it emerged that the separation of powers doctrine can hugely affect the independence and discretion of the enforcement institutions and avoid the formation of conflicts of interests between the regulator and the regulated. Enforcement institutions' independence and the freedom to take discretionary decisions during enforcement activities can, therefore, have a significant impact on the level of criminalisation of OHS crimes. More institutional independence and discretion, such as in Britain, more likely that the institutions will use more compliance-driven policies, especially when funding is scarce. Less institutional independence and discretion, such in Italy, less likely that the enforcement institution will be allowed to use compliance-driven enforcement policies, but there must be an underpinning legal obligation to these enforcement strategies, which in Italy is represented through constitutional principles. The separation of deterrence- and compliance-driven enforcement policies at two different times of the enforcement officers' activities or, better, if conducted between two different institutions, increases the guarantees that the enforcement institution and officers will not be captured by the regulated community's interests and will decrease the chances of the under-, or over-, criminalisation of OHS crimes.

In Britain, the HSE's Enforcement Policy Statement and EMM allow the regulatory institutions to take more discretionary decisions. Hence, enforcement institution can change their enforcement policies and decide, for example, to provide more information and assistance to duty holders during proactive inspections. This freedom is caused by the inexplicit nature of the British politics, which allows state's institutions to continually mediate their powers and activities according to contextual historical needs. This injects flexibility in governance and can improve institutional responses to demands from social factions, but the system is not designed to guarantee a constant and equal redistribution of civil rights. The inequality caused by the enforcement institution's under-criminalisation of OHS crimes is an example. In other words, British citizens' rights are not strictly guaranteed by the state, as they are under the Italian Constitution, but are dependent on political struggles. Hence, law enforcement institutions can mediate the level and stringency of the enforcement policies used with stakeholders, including citizens and the government, but, by doing so, the system might increase the level of social inequality. That is because the mediating processes or political struggles can be greatly influenced by sociopolitical and economic contexts. For example, the acceptance of neoliberal economic policies as the main global economic paradigm is having a significant impact on jurisdictions such as Britain, where the level of civil rights and justice are mediated rather than guaranteed.

In Italy, the CPP and UPG law enforcement regulations do not give institutions and officers the power to use the same level of discretion as their British colleagues. At the moment, the CPP and UPG law give more emphasis to deterrence-driven policies because, Italian officers argued, deterrence-driven policies are more effective when enforcing the law. The Italian Constitution, which regulates state–citizen relationship quite strictly, makes the state automatically accountable to enforce citizens' civil rights, such as ensuring the equality before the law principle to anyone in the Italian territory. Thus, the Italian Constitution is a guarantor of citizens' rights and freedoms and the state's institutions are responsible, and thus accountable, to ensure these rights and freedoms to citizens across social factions. These guarantees are also mediated and can also result from social struggles, such as in Britain, but the mediating process can be slower and reliant on the pluralist nature of the Italian Parliamentary-centred political system. Hence, the reason Italian OHS enforcement policies reformed to become more deterrence-driven is because of Italian constitutional values, which encourage and guarantee social equality and justice.

The enforcement institution and officers' discretion to issue on-the-spot penalties or charge duty holders is an issue that both British and Italian enforcement officers gave importance to. On the one hand, British enforcement officers valued the power to issue on-the-spot administrative penalties, but the reason why this policy is seen as a potential threat is because these would not be regulated by the legislative and checked by the judiciary, such is the case in Italy, but left to the discretion of the enforcement institutions. The introduction of the FFI policy in 2012 means that there is no need to tighten the legislative and Judiciary's control over the HSE, but firms are subject to a higher level of deterrence than they used to be, which improves levels of compliance. The FFI policy has, thus, allowed enforcement officers to maintain their discretion, but also empowered them with a tool deterring firms from breaching the OHS legislation, penalising lawbreakers and promoting a regulatory compliance culture. In other words, the FFI charges represent a rare example of deterrence-driven enforcement policy introduced since the early 2000s in Britain and has, indeed, changed the nature of the enforcement institutions.

The recipient of the fees or penalties paid by the duty holders is also another issue. While in Britain most of the money collected through the FFI policy is paid to the enforcement institution, in Italy the money collected in penalties goes to the Italian Treasury. In comparison to the British FFI policy, Italian OHS penalties cannot be driven by the enforcement institution's need to raise funds. It might be driven by the government's need to raise funds, but these reforms must also be approved in Parliament, which ensures a more pluralistic decision-making process. Hence, the British FFI charging system, on the one hand might incentivise the enforcement institutions to modulate intervention charges on the level of compliance, but, on the other hand, it might cause abuse of power.

British enforcement officers complained that they fear becoming tick-box law enforcers if they were to start issuing on-the-spot penalties or if they were banned from using discretion. This was partially mentioned by Italian enforcement officers, but mostly because they lost the discretionary power they had. They argued in 1994 that they feared the enforcement policies reforms would reduce their roles to mere administrators, but this has not been the case. They lost their enforcement discretion, but acquired the power to issue on-the-spot penalties, and improvement and prohibition notices. Enforcement officers did not associate the new penalties with a form of automated tick-box enforcement approach. Therefore, there is no evidence suggesting that introducing more deterrence-driven policies transforms enforcement officers into tick-box law enforcers.

The reason for the HSE and the government preferring to charge fees rather than on-the-spot penalties is, arguably, to maintain its institutional discretionary independence from Judiciary's scrutiny and Parliament. Both the HSE and the government have a vested interest to maintain the independence of the OHS enforcement activities. That is because the HSE has more control over its operations, and because the government can control those operations without the interference of Parliament, primarily through the control of its funding. Hence, the HSE can use its independence as a bargaining chip while mediating regulatory enforcement activities and funding with the government. For example, it is possible that in 2010 the HSE had to agree to a 35% budget reduction with the government, but managed to bargain the FFI policy in return. This is why one of the British enforcement officers argued in January 2011 that the budget cut would allow the HSE to change the nature of its activities. However, the highest amount of money collected in annual fees so far, £14.9 million in year 2016/2017, is still lower than the overall £84 million annual budget cuts the organisation experienced between 2011 and 2015. The HMRC decision to limit the HSE's FFI revenues collected annually to a maximum of £11 million is not helping the organisation to recover from the recent budget cuts and achieve their statutory scope efficiently. This also demonstrates how the HSE independence from legislative is fictional, and, hence, political pluralist control, but not so from the government executive control.

Thus, the relationship between allowing the enforcement institutions to use discretion and the FFI policy is controversial. British OHS enforcement institutions can use discretion, but since 2012 they had no penalising power unless they were taking employers to court, and won. The HSE enforcement institutions, however, were adamant about starting prosecutions because of their costs and because they would have lost control over the outcomes of cases. Hence, the enforcement institution's discretionary powers were, effectively, decriminalising OHS crimes, and this was especially the case if their budget was reduced. Since 2012, the

HSE has been allowed to start charging fees to employers when found in material breach of the regulation and, effectively, penalising duty holders with on-the-spot charges. This has improved deterrence but has, however, raised a series of questions in terms of the conflicts of interests that the HSE is faced with when charging employers with fees. It questioned whether the HSE charges duty holders more to achieve compliance or to increase its revenues. As a consequence, in 2015, HM Revenue and Customs has limited the amount of money collectable in fees by the HSE to £11 million in order to solve the issue. Hence, the issue is controversial because since the early 2000s the enforcement policies used by the HSE have not been an issue when they were decriminalising duty holders, but have started to become an issue when the institution attempted to increase the criminalisation of OHS crimes.

8

Conclusion

This comparative analysis of the policies and practices of occupational health and safety (OHS) law enforcement in Britain and Italy enable the drawing of several conclusions. Occupational safety incidents and occupational health-related sicknesses are one of the most common causes of suffering in developed countries. Despite this issue, OHS crimes do not attract the same level of attention from criminal justice system institutions as other similarly harmful crimes. This represents an inconsistent reaction of the criminal justice system and, thus since the 1970s, some scholars have argued that OHS crimes are subject to a conventionalised form of under-criminalisation (Carson 1970a, b, 1979; Fooks 2008; CCA 2008). This means that despite the legal responsibility of employers in Britain and Italy to ensure the OHS of workers, these crimes still result in low penalties, are rarely prosecuted, or guilty sentences are rarely achieved at the end of court trials. In addition to that, in Britain the cost of OHS incidents and sicknesses is still mostly paid by workers and their families. This means that if the OHS regulation is defined as a pragmatic response to social problems caused by an industrialisation process driven by a capitalist means of production, it can be argued that the regulation fails to achieve its intended

© The Author(s) 2019
D. Canciani, *The Politics and Practice of Occupational Health and Safety Law Enforcement*, Critical Criminological Perspectives,
https://doi.org/10.1007/978-3-319-98509-1_8

goals. In other words, with the under-criminalisation of OHS crimes, states and their criminal justice systems have failed to redistribute the human and social costs and problems caused by a capitalist means of production and failed to achieve social equality and justice. This research study has considered a number of political, social and legal traditions, as well as economic pressures, when attempting to explain why these crimes are under-criminalised and why in some jurisdictions these crimes are allowed to be further under-criminalised.

The achievement of these goals depends on the enforcement policies adopted by the institutions assigned with the statutory duty to enforce economic regulations, such as the OHS regulation on which this comparative research study focuses. This comparative analysis has used the term compliance-driven to describe all those enforcement activities that focus on the primary goal of achieving compliance by helping and supporting businesses, and deterrence-driven to describe all those enforcement activities that focus on the primary goal of achieving compliance by penalising and prosecuting duty holders. An issue that has exacerbated the under-criminalisation of regulatory crimes is that since the 1970s innovative OHS compliance-driven regulatory enforcement strategies and practices have been theorised by scholars and adopted in a number of jurisdictions. These innovative policies have increasingly stressed the importance of achieving regulatory compliance by supporting businesses with information and other assistive strategies but have, in fact, ignored the under-criminalisation issue. The research studies and theories justifying these innovative enforcement policies have been based on the dramaturgical approaches and ethnographic analyses of the micro-interactions between regulatory enforcement officers and stakeholders (Hawkins 2002). These innovative policies have, thus, failed to analyse critically the social, political, legal and economic issues causing under-criminalisation and injustice. The reason for changing the theoretical framework used to research and analyse this subject, therefore, is the focus on the issue of under-criminalisation and the achievement of the regulatory goals of equality and justice, rather than just the achievement of regulatory compliance.

Indeed, another important aspect of the regulation is how this is designed. Since the 1970s, innovative principle-based economic regulations have gradually substituted standard-based ones (Braithwaite 1987; Hutter 1997; Dubini 2001). This has happened because principle-based regulations are statutes capable of organically adapting to innovative economic technologies and practices without the constant involvement of policymakers. However, this type of regulation requires duty holders to achieve the main regulatory goal, the OHS of workers, as far as this is *reasonably practicable* in Britain, and *technologically viable* in Italy and under the European Union (EU) regulations. The nature of the words *reasonably* and *viable* is already challenging to interpret. However, while the technological viability to achieve the regulatory goal depends mostly on the technologies available, the reasonable practicability of achieving regulatory goals depends also on the funding required to achieve it (Edwards v. National Coal Board 1949). In other words, the British regulation gives to duty holders an absolute responsibility to ensure the health and safety of workers, but only if reasonable and practicable. This can be confusing, especially during a litigation process which involves the regulator or during court trials. The European Court of Justice (ECJ) challenged the British Government on the use of the term reasonable practicability in their OHS regulation, but the legal challenge was rejected after ten years. Arguably, the main reason for this is because reasonability is a principle embedded into the British legal system, which means that an ECJ victory of the case would have set in motion a series of reformation process that could have profoundly changed the British legal system. Arguably, this was out of question. Hence, a contributing factor in the under-criminalisation of OHS crime might be the use of principle-based regulations, and legal terms such as reasonable practicability and technologically viability. This research study has identified also other similar legal and political factors.

Another important legal aspect that might cause the under-criminalisation of OHS crimes is the level of discretion that law enforcement institutions and officers can use while enforcing the regulation. Enforcement discretion is defined as the freedom of the OHS institutions and officers to decide what should be done in particular situations during the course of their enforcement activities (adapted from the English Oxford Dictionary 2017). The literature on the subject identify legal discretion as an essential tool in regulatory enforcement (Bardach and Kagan 1982; Braithwaite 1987; Hawkins 1984; Hutter 1997; Black 2001; Dworkin 1977; Galligan 1986). While enforcement discretion is allowed and embedded in the British criminal justice system, in Italy this is forbidden. The use of discretion during law enforcement activities poses a fundamental challenge to the achievement of the modern principle of equality before the law. This principle is based on the presumption that similar harmful deviant acts should attract similar responses from the criminal justice system. The under-criminalisation of OHS crimes conclusively demonstrates that these deviant acts are not dealt with by the criminal justice system in a consistent way when compared to other similar harmful behaviours. This means that the embedded discretionary practices characteristics of the British criminal justice system give origin to social inequality and injustice.

Another factor that might be contributing to the under-criminalisation of OHS crimes is the principles and values of the legal systems. This comparative research study has considered the crime-control and adversarial tradition of the British legal system and due-process non-adversarial approach of the Italian one. Both legal systems aim to achieve legal efficiency, which is 'the system's capacity to apprehend, try, convict, and dispose of a high proportion of criminal offenders whose offences become known' (Packer 1968, p. 158). However, these two legal systems aim to achieve this legal efficiency in different ways. The crime-control adversarial system, traditional in Britain, has been compared to a conveyor belt; the due-process non-adversarial one, traditional in Italy, to an obstacle course. The crime-control adversarial system focuses on ensuring the rights and liberties of law-abiding citizens, which means that the criminal justice system gives much more importance to the apprehension of criminals. In this legal system the accused guilty is presumed from the start. The due-process non-adversarial system has traditionally been focused on ensuring the rights of all citizens, which means that there is much more emphasis placed on the civil rights and freedom of suspected criminals. Thus, the accused must not be presumed guilty by state institutions during the adjudication process (Packer 1968; Damaška 1973, 1986; Sanders and Young 2007).

In the crime-control adversarial criminal justice system the trial of criminals must be conducted quickly so the accused, victims and witnesses are able to produce more accurate accounts of the incident, which improves the quality of the evidence and helps to achieve the fairer outcome. This system also considers the financial costs of criminal procedures, which means that it welcomes fast-track trials, guilty pleas, and less regulated procedural practices. The due-process non-adversarial system has traditionally been more concerned with finding the truth, which requires more time and, hence, is more expensive. Guilty pleas and fast-track trials might abuse the accused's rights to a fair trial, which means that the due-process non-adversarial system is more regulated by enforcement procedural rules and regulations (Packer 1968; Damaŝka 1973, 1986; Sanders and Young 2007).

In the crime-control adversarial legal system, prosecutions have traditionally been commenced by law enforcement institutions, which are given more discretionary powers on whether to lay charges against the accused. Hence, in Britain, for example, the enforcement institution's decision to start a trial also depends on the public interest test, which can consider a wide range of social values and principles. During court trials, judges have the role of referees and cannot interfere in collection of criminal evidence. In the due-process non-adversarial system law enforcement institutions have less discretionary powers and the decisions to proceed to trial depends on the quantity of criminal evidence available (Packer 1968; Damaŝka 1973, 1986; Sanders and Young 2007). These, however, are assessed by an institution solely responsible to prosecute criminal offences, which in Italy is the Public Ministry (PM). The criminal justice system institutions are regulated through strict procedural processes, such as the Italian Code of Penal Procedure (CPP), which significantly reduces the level of procedural discretion. For example, Italian prosecutors must follow the principle of legal obligation, which means that they must start a trial if there is enough criminal evidence available. In this system, judges are active participators in the adjudication process and can, for example, require the collection of more criminal evidence (Gustapane 2012).

Another major factor that might be contributing to the under-criminalisation of OHS crime is the separation of powers doctrine, which is the way the state's institutions powers are regulated, including the relationship that these have with citizens. This is important for this comparative research study because the way the OHS enforcement institutions are controlled will also influence the enforcement policies they can use. Secondat and Montesquieu's (de Secondat 2001) theory of the separation of power doctrine asserts that to avoid despotic rulers, guarantee personal civil liberties and create democratic governments the state should be formed of three main institutions, each with specific responsibilities; parliaments should enact legislation, governments propose and execute them, and the judiciary judge legal violations. These systems have developed from historical traditions and both the British and Italian political systems broadly conform to these principles (Vile 1963; Langbein et al. 2009; Mousourakis 2015). The difference between Britain and Italy, however, is that in the former the government has much more powers than the parliament. This executive authority comes from the historical royal power, which has gradually been transferred to the cabinet. The Italian Parliament has more power than the government, which can only execute legislations agreed collectively. Hence, for example, the British Health and Safety Executive (HSE) enforcement policies and annual budget are decided by the government, while in Italy the Prevention Services for Safety in Work Environments's (SPSAL) enforcement policies and budget are decided by the Parliament through a more pluralist decision-making process (Warwick 2006).

In fact, the way OHS enforcement resources are decided might also be one of the major causes of the under-criminalisation of OHS crimes. While the British Government has an extended amount of power to decide the budget of the HSE, in Italy the SPSALs budget is decided by Parliament through a collegial decision process. In addition, the rationale by which resources are decided in the two jurisdictions differs. Due to constitutional values and principles, in Italy OHS resources allocated for deterrence-driven enforcement activities cannot fall under a certain level. This is not the case in Britain where government decisions on the matter are not limited by the same principles. Hence, while in Britain the enforcement policies and level of criminalisation of OHS crimes resulting from the operation of the enforcement institutions depend also on the resources enforcement institutions receives from the government, in Italy it is enforcement policies that the enforcement institutions must use that determine the level of resources that they need to receive. In other words, the system adopted in Italy to decide the resources dedicated to the OHS regulatory enforcement institutions ensure a consistent minimum level of deterrence-based enforcement activities and, hence, a more constant level of criminalisation of these crimes.

A comparative analysis of British and Italian OHS enforcement policies offers a significant contribution to the subject because the two jurisdictions have, to some extent, opposite enforcement strategies. That is because this comparative analysis also considers the social, political and judicial reasons, arguments and historical reformation processes that have led to these two different regulatory regimes. These two jurisdictions, which are both still part of the EU at the time of writing, have similar regulatory frameworks due to the 1989 European Economic Community (EEC) Council Directive on health and safety (89/391/EEC), but have different enforcement strategies and policies. Hence, to understand and explain the reasons for these different regulatory enforcement strategies and attempt to understand how and why OHS crimes are under-criminalised, this comparative analysis has adopted a critical approach by scrutinising these two jurisdictions' political and legal systems and traditions. Cross-national comparative analyses are extremely challenging and the differences between these two countries' social, historical and political traditions can barely be accounted for in this book, but as Nelken (2010) argues, these differences are exactly the reasons why these comparisons are worth pursuing.

1 History and Context

Occupational OHS regulation have been introduced during the various industrialisation processes that occurred across jurisdictions in Europe in the nineteenth and twentieth centuries. In both Britain and Italy, the level of regulation and the enforcement institutions responsible to enforce the law have continued to grow in size and scope. The regulation and the enforcement institutions, in other word, have grown to prevent the social problems caused by a capitalist mode of production and, thus, as an economic redistribution means which ensures justice and equality (Bartrip and Fenn 1980a, b, 1983; Taylor 1972 as cited by Ogus 1994; Slapper 2000). In 1974, the first OHS goal-setting regulation was introduced in Britain, which regulates the outcomes, rather than the processes, that duty holders should achieve by law. This new philosophy was adopted across Europe with the ECC Directive 89/391/ECC, and in Italy introduced with the D.Leg.vo 626 1994 (Repubblica Italiana 1994; European Commission 2007, 2012; Delle Fave 2013; Fioravanti 2011; Council of Europe 2010).

The organisation and responsibilities of the enforcement authorities in both jurisdictions differ. The main British enforcement institution is the HSE, a nationwide Executive Non-Departmental Public Body associated to the DWP, which is accountable and responds to the Secretary of State for Work and Pensions (SoSWP). Local authorities (LAs) also have OHS enforcement responsibilities, but mainly towards the service sector. LAs are expected to follow HSE enforcement guidelines decided by the HSE and SoSWP through liaisons committees and panels. Both organisations are very financially efficient in enforcing the regulation with deterrence-driven and compliance-driven policies. The HSE claimed independence from the government is limited due to the SoSWP's involvement in organisational decisions. This institutional organisation is very cost-effective, but raises concerns in terms of regulatory capture and the conflict of interests that enforcement officers might encounter during their field activities. In 1980, the SPSALs, organisations subordinate to the Ministry of Health and regional health services, became the main OHS enforcement authority in Italy. Labour Territorial Inspectorates (ITL) institutions, which are subordinate to the Ministry of Work and Social Policies, have traditionally been responsible for enforcing labour laws, including OHS, but since 1980, their OHS enforcement role has been marginalised to a few sectors of the economy. Both enforcement institutions must coordinate enforcement activities jointly and in accordance with the Penal Code (CP), the CPP, and as regulated by the UPG (Pais 2008; Rinaldi 2012). Hence, Italian Ministers have less power to affect OHS policies and strategies then the SoSWP in Britain.

An additional key OHS institution in Italy is the National Institute for Insurance against Working Accidents (INAIL). This is the national compulsory OHS insurance administration for businesses and workers. The insurance scheme compensates workers for OHS incidents, which also includes the payment of the workers' wage. Thus, INAIL is also responsible for collecting incidents statistics. Contrary to the British statistics, which Tombs and Whyte (2008) calculate to be subject to mis- and under-reporting by a factor of five to six times the official HSE figures for non-fatal incidents, Italian statistics seems to be much more accurate in counting incident statistics. This is because the insurance compensation scheme incentivises both workers and employers to report incidents. Since the early 2000s, INAIL has increasingly acquired a regulatory promotional role by distributing grants to SMEs and aiming at improving OHS in the economy. Since 2013, INAIL overall annual grants for SMEs have averaged €300 million. Hence, INAIL has taken over the compliance-driven enforcement policies that SPSALs were implementing between 1980 and 1994 (Pais 2008; Rinaldi 2012). This has significantly reduced the conflict of interest that SPSALs officers experienced (Rausei 2006; Porreca 2008).

Since the 1970s, innovative forms of regulatory enforcement policies have been theorised and adopted by a number of jurisdictions, but these innovative strategies have neglected to resolve the problem of the under-criminalisation of OHS crimes (Carson 1979). Bartrip and Fenn (1980a) argue that this form of under-criminalisation had been happening since the mid-nineteenth century, when the British Factory Inspectorate, which, due to the lack of resources and magistrates' conflicts of interest, had to shift enforcement policies from deterrence-driven to compliance-driven in order to achieve some results and improve regulatory compliance (Bartrip and Fenn 1980a, b, 1983). Until the end of the 1990s, in both Britain and Italy the growth of enforcement institutions followed similar patterns. Since then, however, the expansion and scope of the OHS enforcement institutions in Britain has stopped and, perhaps, reversed, while in Italy it has continued. Since the early 2000s, the British OHS enforcement policies have been reformed to become even more compliance-driven than ever, while in Italy these were reformed into deterrence-driven policies since 1994.

Since 2004, the HSE enforcement officers have witnessed a constant reduction in their capabilities to enforce the regulation with deterrence-driven policies. These reforms have been imposed to the regulator by reducing the level of funding available to the institution and by suggesting the use compliance-driven policies, despite widespread suggestions from a stake-holder's consultation in 2000 that more deterrence-driven policies were needed to improve compliance level further. Compliance-driven policies have been introduced because they are more financially efficient to implement than the more traditional deterrence-driven ones, and are capable of reliev-ing businesses from the burden of regulation while being helped to achieve compliance. The 2008 global financial crisis has given British governments a renewed justification to start political campaigns arguing for reduced fund-ing to public institutions, both to save money in times of economic crisis, and give businesses space to re-boost economic output without the burden caused by regulation and regulatory enforcement. By 2011, the newly elected British government renewed the promise to reduce the regulatory burden of businesses and decided to cut the HSE budget by a further 30% by 2015. Despite the introduction of the Fee for Intervention (FFI) policy, which con-tributes at most by £11 million to the HSE yearly budget, in real terms and at purchasing power parity (PPP), the HSE's annual budget has been reduced by 48.8% between 2000 and 2016. This has eroded enforcement activities, further decriminalised OHS crimes and increased social injustice.

In Italy there have also been some attempts to reduce the regulatory burden of regulation on businesses, but these have not been as effec-tive as in Britain. Due to constitutional values, which are designed to promote and, hence, guarantee citizens' rights, the decriminalisation of OHS crimes has been much more limited. Three main reasons can be identified. First, the Italian Constitution guarantees a minimum level of law enforcement, including that of OHS. Visiting 5% of firms annu-ally is considered the minimum level of proactive enforcement activi-ties. Second, the Constitution also imposes a strict separation of power between the state executive, legislative and judiciary. When compared to the British political model, the Italian constitutional-driven separa-tion of powers doctrine gives much more authority to the Parliament to decide whether to pass legislation, reduce public funding, or change OHS enforcement policies. Hence, the more pluralist nature of Italian politics, and a strict separation of powers, has prevented a radical ref-ormation of the OHS enforcement policies. Third, the traditional design of the Italian criminal justice system, which abides strictly to due-process principles of legal consistency and proportionality, has also prevented this decriminalisation process. In comparison to Britain, the procedures dictating the decisions to prosecute OHS crimes must abide by the CP and CPP. These procedural codes are designed to achieve similar response to criminal acts across the whole spectrum of crimes, not only OHS. The CP and CPP attempt to match the social harm caused by criminalised acts to the responding punishment of the criminal justice system and the procedures by which the investigation and prosecution might start. This is not the case in Britain, and less so in England, where criminal justice system responses to crimes, albeit increasingly rationalised by the Crown Prosecution Service (CPS), are not standardised across *all* deviant actions. In other words, criminal justice system responses to crimes in Britain lack consistency.

The level and types of penalties issues to non-compliance also differ radically between Britain and Italy. British OHS enforcement officers did not have the power to issue any on-the-spot penal charges to non-compliers until 2012 when, for the first time in history, they were given the power to charge duty holders for the technical support provided while encountering material (i.e. significant) breaches of the regulation. The new FFI charging system transfers the cost of non-compliance to non-compliant duty holders and firms, which is a welcomed deterrence-driven policy. However, in 2015, the HMRC forced the HSE to pay any revenue collected from the FFI policy above £11 million to the Treasury. This is not the case in Italy where SPSALs officers have been issuing on-the-spot fixed *compulsory* penalties since 1994. In Italy, penalties are decided in the Parliament, monitored at arm's length by the PM, and collected by the Treasury (Rinaldi 2012).

The most interesting differences between the two jurisdictions is the reasons why judicial on-the-spot administrative penalties cannot be issued in Britain but represent a standardised routine in Italy, and why British enforcement officers can charge firms for providing support, but Italian jurists consider such act a profanation of the fundamental principles of the Italian criminal justice system. According to enforcement officers, the British (and English in particular) Judiciary perceives any on-the-spot administrative penalty issued by the enforcement institutions as a threat to citizens' liberties and an abuse of power by the government. The HSE FFI policy was introduced to obviate the lack of on-the-spot penalties and persuade duty holders to actively seek compliance *before* officers' inspections, rather than only after. This increased the efficiency of the enforcement activities because compliance was achieved in advance by deterring duty holders. Italian law enforcement agencies cannot charge duty holders for their services because it would question the SPSALs' institutional independence and create conflicts of interest between the enforcement officers' primary objective, which is to achieve greater compliance among duty holders, and the criminal justice system's main objective of achieving justice. The Italian OHS enforcement institutions' organisation and powers, in other words, are designed to avoid this conflict and achieve social justice consistently, proportionately and, hence, fairly (Rausei 2006, 2008). This is not the case in Britain, and particularly in England, because the penalties issued are not strictly regulated by an Italian equivalent of the CP, and enforcement policies are also decided by the SoSWP, who is a cabinet member.

British and Italian enforcement institution's decisions and procedures to prosecute also change significantly, but nevertheless OHS crimes are under-criminalised in both jurisdictions. In Britain the decision to prosecute is taken by the enforcement institution as guided mainly by principles of the Code of Crown Prosecution, which are the availability of evidence, a realistic prospect of conviction and the public interest test (HSE 2015). In Italy, according to the CPP, cases classified as prosecutable are automatically referred to public prosecutors, whose decisions must abide by the principle of legal obligation and, hence, occur automatically if there is enough evidence of criminal conduct. The government has neither direct political control influence over these decisions, nor indirect, such as through the funding to be allocated to the enforcement institution. Also, HSE and LAs are responsible for taking cases to court directly, without transferring cases to the police or the CPS (except in particular situations, such as for manslaughter cases), which means that the institution is also responsible for funding court trials. Hence, the HSE has a strong interest to maximise the prosecution rate of OHS crimes in order to use resources efficiently. In Italy, this cannot happen at enforcement institution stage, but prosecution statistics are not available to analyse whether OHS court cases are more frequent than in Britain and what is the rate of prosecution achieving guilty sentences. However, according to an Italian enforcement officer, the rate of prosecutions achieving conviction is about 50% as compared to the more than 90% in Britain. The issues in Italy seem to be mostly caused by the incompetency of public prosecutors and judges in understanding OHS technical matters and a lack of resources needed to resolve this issue. Hence, while in Britain under-criminalisation is caused by the enforcement institutions practices and policies, in Italy it happens at court trial stage.

2   Incidents and Enforcement Trends

Another major dearth in the literature proposing the introduction of compliance-driven OHS enforcement policies is the issue of the level of harm that these social phenomena cause. Unsafe working conditions and the occupational health issues resulting from it are one of the causes of death in developed countries. However, this seems to be rarely accounted for in official statistics (Tombs and Whyte 2008). One of the significant issues in Britain is that the official statistics omits to count for non-fatal incidents by a factor of five to six times. Arguably, this underestimation has allowed British governments in recent years to justify the introduction of enforcement policies reforms further decriminalising OHS crimes, or to justify a reduction of the enforcement institutions' budgets. Hence, this research project attempts to measure and compare the fatal and non-fatal incidents recorded in Britain and Italy. The comparative analysis between these two jurisdictions also attempts to account for the under-recording of non-fatal incidents in Britain.

In both Britain and Italy incident levels have fallen steadily in the past decades. However, the improving trends in the two jurisdictions have followed different paths. While in the former most of the improvement occurred up to the end of the 1990s and slowed down after that; in the latter incident trends improved slowly until the end of the 1990s and accelerated after that. An important observation emerging from the comparison is that after the British fatal incidents' figures are adjusted to include work-related road incidents, and the non-fatal ones to include the mis- and under-reported data, as estimated by Tombs and Whyte

(2008), the British fatal and non-fatal incidents per 100,000 workers appear to be higher in Britain than in Italy.

This is not to say that in Italy there are not under-reported incidents, but due to the INAIL compulsory insurance scheme, all Italian stakeholders are incentivised to report OHS incidents. That is because the OHS compulsory insurance scheme run by INAIL covers for incidents' costs and the injured workers' wages. Workers also have an incentive to report incidents because they receive 100% of their wage, as compared to 80% when on sick leave. Hence, Italian OHS incident figures might not be subject to the same level of under-reporting of the British one. This is a critical observation because the claim that until the 2005 EU enlargement, Italy held the worst OHS records might not be accurate, and this might still be the case in the present day when analysing British, but also other jurisdictions' OHS incident trends.

These trends become notable also when looking at the OHS enforcement policy reforms implemented in Britain and Italy. In Britain, the compliance-driven enforcement policies widely adopted from the early 2000s have not managed to decrease OHS incidents significantly. In Italy, the opposite happened. A sharper fall of OHS incidents followed the mid-1990s regulatory law enforcement policy reforms, which injected deterrence-driven practices in SPSALs' enforcement policies. In other words, this comparative quantitative descriptive analysis of OHS incidents and enforcement policies suggests that deterrence-driven enforcement policies, backed up by on-the-spot penalties, are more effective in preventive incidents, achieving compliance and reducing the under-criminalisation of these crimes.

3   How Enforcement Resources Affect Enforcement Policies

Another notable dearth in the literature supporting the use of compliance-driven enforcement policies is the lack of association between resources and the enforcement institutions' policies used. Tombs and Whyte (2010) link the consistent fall in resources that occurred in Britain with its effect on the volume and extent of enforcement policies adopted by the British regulator. In addition, Bartrip and Fenn (1980a, b, 1983) have demonstrated that the British Factory Inspectorate's reasons for switching to compliance-driven policies by the 1850s happened due a lack of resources to effectively use deterrence-driven policies and conflict of interests between the regulation and magistrates. HSE enforcement policies enacted by government, which are not subject to a strict separation of powers doctrine, and are weakly designed to avoid conflicts of interests, have the potential to generate great social harm and injustice. In Italy, no one has ever analysed the enforcement statistics available or the opinions of the enforcement officers on resources, but according to Italian constitutional values the state must guarantee enough resources to ensure a minimum level of deterrence-driven enforcement activities across all law enforcement agencies.

Financial resources have fallen in Britain by 47% between the financial years 1999/2000 and 2016/2017. Data for LAs also show a steady decrease in resources for OHS enforcement institutions. Italian statistics suggest that while SPSALs resources have remained stable in the 2000s, INAIL has distributed funds through grants to SMEs for projects aimed at improving OHS. The scheme was piloted in the early 2000s, and by 2011, SMEs had access to roughly €300 million per year in grants. While in Britain since the early 2000s resources have been cut throughout the OHS enforcement institutions, in Italy resources have been injected into schemes aimed at helping SMEs improve their compliance levels. This has caused British enforcement institutions to decrease the level of deterrence-driven enforcement policies, while in Italy the deterrence-driven enforcement policies have not decreased but there has been a renewed emphasis on compliance-driven policies through INAIL.

From the interviews it emerged that there is a consensus among British and Italian OHS enforcement officers that low levels of funding and human resources affect their field operations negatively and put their enforcement activities under pressure. The resources to conduct what British and Italian officers consider appropriate OHS enforcement activities to curb the harm caused by OHS are scarce. All interviewees expressed a level of distress about their workload and the pressure to perform adequately with the resources available. This was the case for both the British and Italian enforcement officers. A British HSE officer argued in astonishment that stress-related leaves among HSE enforcement officers had increased in 2011, which was the year after when the government announced an HSE budget reduction of 30%. Most interviewees in both countries complained that lack of resources resulted in the ineffective performance of enforcement activities. However, most of the British and Italian enforcement officers did not see resources as being a political issue. Only one Italian and one British officer argued that lack of funding is a political decision.

Another interesting finding that emerged from the interviews is that both British and Italian enforcement officers engage in calculations of the costs and benefits of the enforcement policies used, either intentionally or unintentionally. The level of discretion enforcement institutions and officers are allowed to use is a key determinant in this context. In Britain, the discretion embedded in enforcement policies used means that enforcement officers and institutions can engage in cost-benefit analyses (CBAs) and might adapt the enforcement policies to suit the level of resources available. For example, since the early 2000s the HSE has decreased the number of cases taken to court for prosecution in order to improve their prosecution success rates and use resources more efficiently. In Italy the prohibition of using discretionary practices and the strict legal framework within which the institution and officers must operate means that they have almost no power to adapt practices to the resources available. Hence, from this qualitative analysis it can be argued that lack of resources together with enforcement institution discretionary practices, such is the case in Britain, can lead to the use of more financially efficient enforcement policies; however, it can also lead to their decline, and cause the under-criminalisation of OHS crimes. This issue seems to be affecting most regulatory frameworks, and mostly those regulating social elites. Minimum guaranteed level of OHS enforcement activities, together with non-discretionary enforcement practices, such is the case in Italy, means that the Parliament cannot reduce resources below the minimum, which has been 5% of active firms for a long time. Hence, the Italian system ensures constant protection of workers' rights, regardless of other contextual factors, such as the state of the national economy or other international political agreements. This is not the case in Britain.

Enforcement deterrence depends on the likelihood of detection and the level of penalties issued (Becker 1968). While the likelihood of detection can be significantly affected by resources, in the context of OHS enforcement the level and frequency of the penalties issued can greatly increase the efficiency and legitimacy of the enforcement institutions. The use of on-the-spot penalties, in other words, is an effective enforcement tool to achieve regulatory compliance.

Before the FFI policy was introduced in Britain in 2012, even if the number of British enforcement officers increased, their deterrent effect would have not improved, and only partially in cases when they were issuing improvement or prohibition notices. The same applied to their Italian colleagues between 1980 and 1994. Since 1994, breaches in Italy have been penalised by compulsory administrative on-the-spot penalties and the generally stable level of resources for field operation means that the enforcement institution has managed to radically increase its deterrence power at little or no extra cost. Enforcement penalties, or fees, are a quick and financially efficient means to ensure compliance because employers have an incentive to comply with the regulation *before* inspections, and officers can prevent the occurrence of incidents with minimum expense and an immediate return in terms of regulatory compliance.

Thus, deterrence created with low but frequent on-the-spot financial penalties or fees is one of the most efficient and effective enforcement strategies to ensure compliance among duty holders. One of the major issues with penalties, both on-the-spot and issued after court trials, is that the deterrent trap. During proceedings, if firms threaten to declare bankruptcy due to the high level of penalties, and employees are made redundant as a consequence, the regulation will have a detrimental effect on the individuals that is meant to protect. On-the-spot penalties used for minor breaches, therefore, should be low but frequent because they would need to be suitable for both SMEs and larger organisations, and hence reduce the discrimination between these two types of firms. Enforcement charges, such is used in Britain also work, but to avoid conflict of interests the fees collected should be sent to the Treasury rather than be collected by enforcement institution. This strategy causes deterrence and achieves compliance before enforcement officers' inspections, but more frequent enforcement activities also increases enforcement costs. Penalties issued after trials, following serious breaches or incidents, should be higher because they need to have a higher deterrent power over firms. The idea is that the social harm caused by serious offences and incidents is greater than that caused by making workers redundant after dissolving firms with bad OHS records.

Resources also affect the number of prosecutions conducted, as well as the rate of guilty sentences reached. The way prosecution cases are selected for prosecution in both countries is also significant to determine the level of criminalisation of OHS crimes. In the former the OHS enforcement institutions have discretion to decide whether to take prosecutable cases to court and are not required, with the exception of manslaughter cases, to refer cases to the police or CPS. In the latter, penalties and prosecutions are compulsory, and public prosecutors cannot refuse to start a prosecution case unless the evidence of criminal actions available is judged insufficient. Hence, the main difference between Britain and Italy is that while in the former a lack of resources determines the under-criminalisation of OHS crime at enforcement institutions level, in the latter, lack of resources causes their under-criminalisation at prosecution stage. In other words, in Britain lack of resources prevents the enforcement institutions from taking cases to court, while in Italy, under-criminalisation is caused by a lack of funds available to the Judiciary Authority and the PM, as well as public prosecutors' and judges' lack of technical expertise on the subject, which is essential to reach guilty sentences. Hence, in Britain, the OHS enforcement institution's budget should guarantee a minimum and constant level deterrence-drive enforcement policies, while in Italy public prosecutors and judges should be trained to understand the complexities and technicalities of OHS cases.

In Britain, Packer's (1968, p. 158) definition of legal efficiency, 'the system's capacity to apprehend, try, convict, and dispose of a high proportion of criminal offenders', is mistaken for the more basic definition of financial efficiency. Legal efficiency, in other words, is not about costs, but achieving justice is expensive. This is recognised and embedded into the British criminal justice system, but not in the Italian one. Again, the literature proposing alternative compliance-driven enforcement policies fails to take into account this fundamental issue. Snider (1991) suggests that nation states using a crime-control model and lacking written and codified constitutions are at risk of witnessing the transformation of the governing classes into political powers serving the interests of the wealthier. The under-criminalisation of OHS crime is a symptom of this issue. This is happening in Britain faster than in Italy, where the Constitution and the Constitutional Court responsible for enforcing it are designed to prioritise citizens' civil liberties and rights, such as justice and equality, over economic surplus. The civil rights guaranteed by the Italian Constitution, however, are ever more challenged by global neoliberal economic policies and other international institutions and agreements, which constantly aim to decrease state involvement in the economy, cut the state expenditures and its role of guarantor of citizens' civil rights. An example of this is the countless constitutional reforms that been proposed in recent years in Italy, such as the latest 2016 Renzi unsuccessful Constitutional reform referendum, which attempted to reduce citizens' democratic decisions.

While Britain's balance sheet has been gradually adapting to the requirement of a global economic cost-cutting neoliberal policies, the obstruction that the Italian Constitution is posing to this plan is being counter-attacked with financial weapons, which are also responsible for causing endless spiralling level of national debt and consequent political crises (Caracciolo 2015; Zagrebelsky 2013). The unregulated nature of the British political system means that the only opposition that this neoliberal paradigm can expect is through trade unions' and other social movements' struggles, which, arguably, political actions are also undermined by their own ontological doubts that there is no alternative to austerity or a capitalist society. In Italy, the Constitution represents another layer of opposition to these reforms, but the subtle, yet monstrous, effect of these changes are threatened by a political class whose decisions are ever more justified by continuous TINA mantras and who see the Constitution as threat for the state survival. The increasing under-criminalisation of OHS crimes, caused by austerity policies in this instance, is a symptom of the erosion of civil rights and fundamental values of social equality and justice.

Hence, the rationale by which resource affect enforcement policies and vice versa in the two jurisdictions is fundamentally different. In Britain, enforcement activities can be changed according to the amount of funding available. In Italy it is the opposite, resources must be adjusted to the enforcement target and policies required to be achieved. The British system is cost-effective, but it is not designed to ensure equality and justice. The Italian system is designed to ensure equality and justice, but it is costly. Either way, OHS crimes are under-criminalised, but only at two different stages of the criminal justice system. Both jurisdictions OHS deterrence-driven enforcement policies are in one way or another.

4   Discretion Fairness or Consistency?

This research project has reached the conclusion that discretion represents a form of power for OHS enforcement institutions and officers. In this comparative analysis it emerged that the British regulator has more decisional power than their Italian counterpart. Discretion can be used to take the best enforcement decisions to achieve regulatory compliance, but it also allows enforcement officers and agencies to take the best decisions for themselves, or the best decision for the institution controlling it. More discretion does not mean necessarily less enforcement, accountability or social justice and equality, but this study concludes that the use of discretion intertwines with different political and criminal justice system traditions and practices and responses, which increases the likelihood of under-enforcement and under-criminalisation of OHS crimes.

Enforcement discretion plays different roles in the crime-control adversarial system and the due-process non-adversarial one. Discretion has a somewhat central role in the crime-control adversarial model of criminal justice systems. That is because it is the presumption of the defendant's guilt that leads to the beginning of the adjudication process. This means that law enforcement institutions are more likely to be independent from the judiciary, which role is to ensure that the law is applied correctly, and citizens' civil rights respected. The enforcement institution is empowered to take independent decisions in order to take reach verdicts faster and be more legally and financially efficient. This is not strictly the case in the due-process non-adversarial criminal justice system, where adjudication process starts with the acknowledgement of criminal conduct by a law enforcement entity. In due-process non-adversarial systems the prosecutor does not presume the guilt of the defendant and legal efficiency is achieved through a close consideration of the evidence available. In the due-process non-adversarial legal systems discretion is limited and increases as legal cases move further into the judiciary's hands.

The more the state is committed to the separation of power doctrine, the less likely is that enforcement institutions will be given discretion and the less likely it is that crimes will be over- or under-criminalised. That is because the more discretion is given to institutions and officers to decide the enforcement policies to adopt, the less control the judiciary and the legislative will have over their operations. The less control over the enforcement institution's decisions from the judiciary, such as the institution's relative freedom to decide the consequences of legal breaches instead of public prosecutors or the CPS, thevmore likely it is that OHS enforcement institutions will act inconsistently. Less procedural control and institutional decisions from parliaments, such as the enforcement institution's independence to decide the type of enforcement policies and frequency of their activities, the more likely it is that the law will be enforced inconsistently, and thus lead to an under- or over-criminalisation of crimes. Given that the HSE is not as directly accountable to the public as other law enforcement institutions, the dearth of democratic control expressed through the independence of the enforcement institution from Parliament, and its intimate relationship with the government, has the potential to generate inequality and injustice. The enforcement institution's independence confers on it the capability to adapt to social and economic trends quickly, which can be beneficial for the institution and to ensure regulatory compliance, but the social costs caused by this seem to be higher than the benefits achieved, especially for less affluent social classes. In Italy, the stronger separation of power imposed on state institutions leads to a more consistent, and hence equal, criminal justice system treatment of citizens from different social classes. Despite this, the capability of the Italian Judiciary Authority, such is the case for the British enforcement institution, to successfully enforce and prosecute OHS crime also depends on other factors.

In both jurisdictions the relative ambiguity of the regulatory objectives of the principles-based OHS legislation means that duty holders might have acquired power to defend their harmful practices better. This is the case at shop floor level, which is more relevant in Britain where discretionary practices allow for more interaction and dialogue between the regulator and stakeholders. However, this is also the case in Italy during court cases. The responsibility of duty holders to ensure the OHS of workers as far as it is reasonably practicable in Britain and technologically viable in Italy also seem to create barriers to apprehend, penalise and convict these crimes more effectively. However, while the principle of technologically viability in Italy allows for an easier determination of duty holders' culpability, the principle of reasonable practicability in Britain seems to be better for defending duty holders' conduct and, thus, their legal liability for OHS crimes. In other words, the principle of legal reasonableness, which is an embedded feature of the British criminal justice system, in the context of OHS crimes has been and currently is facilitating the under-criminalisation of these crimes.

The relationship between discretionary practices and resources is also significant in the context of OHS enforcement policies. If enforcement institutions and officers are given the discretion to change enforcement practices, their practices and the level of criminalisation of crime will be positively correlated to the amount of funding available, such is the case in Britain. The enforcement institution's incapability to adapt enforcement policies to the resources available means that changes to the institution budget might have an impact on enforcement performance level, but the enforcement policies will not change. This is a very important consideration to make in the current socio-economic global climate, where neoliberal economic policies, austerity measures and political rhetoric supporting TINA mantras, are gradually eroding public institutions' capacity to achieve their statutory duties and, hence, preserve citizens' civil rights. Britain is doing great because the discretionary practices embedded in politics and in the criminal justice system enables the reduction of the state's expenses easily, but this comes with a price for society and particularly for the least affluent citizens. The less discretion conferred to enforcement institutions or officers, such is the case in Italy, the less likely under- (or over-) enforcement and criminalisation will occur. However, state institutions are not capable of achieving the same level of financial efficiency because their practices are not discretionary but determined by Constitutional values aiming at preserving citizens' civil rights.

There seems to be a relationship also between institutional accountability and discretion. This relationship depends on a straightforward question: who should control the controllers? The more independent is an institution and the more discretion it can use, the more accountable it should be to stakeholders through direct elections of its governing officers. Arguably, the institution should also be independent of parliament and the government, and collect revenues from stakeholders. This represents the outsourcing of enforcement responsibilities to private organisation, which, however, creates conflicts of interests and potential regulatory failures. An example was the inadequacy of the big four global accountancy companies' capability to report or react to the issues that caused the 2008 global financial crises. Braithwaite (1997) proposes interesting alternatives to improve these relationships but makes the reader wonder whether private institutions are better for this role than democratically controlled public institutions. This depends on whether the state is constantly guaranteeing pluralist decision-making, or whether the state is simply a political institution funded to preserve capitalist interests. If the latter is the case, modern liberal principles of equality and justice will never be implemented fully, and the under-criminalisation of OHS crime will never be solved. In Britain, the HSE is quite closely controlled by the government, even if its legal status as Executive Non-Departmental Public Body, suggests differently. The government holds the institution's purse strings, and Parliament has little say on the resources given or the enforcement policies used. The attempt of various Parliamentary committees to respond to this issue in the past have failed and exposed the lack of accountability the HSE has toward stakeholders. In other words, the British OHS regulator is not publicly accountable enough for the level of decisional discretion it enjoys.

Thus, OHS regulation is enforced more consistently and proportionally in Italy than in Britain. That is because the British legal system consistency is replaced by the concepts of legal fairness and proportionality, which are defined as the achievement of uniform *outcomes*, while in Italy legal consistency ensured uniform *actions*. Achieving uniform regulatory outcomes during OHS enforcement activities means that the actions of the enforcement officers are not consistent to other similarly harmful crimes, and because their enforcement decisions are also adjusted to the typology of the firm and the behaviours of employers. As also observed by one of the interviewees, this compliance-driven approach is very different from those adopted by other law enforcement agencies, such as the police, which usually uses a more deterrence-driven approach. Although Italian OHS crimes are also under-criminalised, this seems to occur mostly at court trial stage, because neither the enforcement institutions, nor the public prosecutor within the PM can refuse to take cases to court if the evidence of criminal conduct is judged sufficient. Thus, in Italy discretionary decisions are mostly taken by member of the Judicial Authority.

5   Discretion and Pragmatic Enforcement Activities

In day-to-day activities, discretion can also cause the under-criminalisation of OHS crimes. The planning of annual proactive enforcement activities can be affected by the level of discretion the enforcement institutions enjoys. The British targeting system is much more advanced and centrally controlled, which means that enforcement officers have less discretion to choose the firms to visit. However, the British proactive enforcement targeting policies do not guarantee a minimum level of proactive activities, such is the case in Italy, which means that if the HSE decrease the annual enforcement inspections targets, they also decriminalise OHS crime. In Italy the OHS enforcement institution inspects 5% of the firms in the territory. However, the specific local enforcement targets are decided by the chief enforcement officers and based on national enforcement targeting guidelines per economic sector.

Italian chief enforcement officers are given a level of discretion to adapt the national enforcement targeting guidelines to the specific economic composition of their local jurisdiction. This discretion, however, can also be used to favour the achievement of annual targets by, for example, choosing firms known to be compliant and, hence, save time and reduce workload. The Italian strategy has, thus, the potential to decrease the criminalisation of OHS crimes and is less centrally controlled and hence accountable than the British targeting strategy.

The level of discretion that enforcement institutions have a significant impact on the enforcement policies they can use, and thus on the level of criminalisation of OHS crimes. The European directive suggests member states to have an enforcement institution designed to enforce the regulation through deterrence-driven strategies and another to help duty holders achieve compliance by providing help and information, that is, through compliance-driven enforcement policies. This is to avoid conflicts of interest. These conflicts of interest, at the moment, are better avoided in Italy than in Britain. This is shown with the reformation of SPSAL's enforcement policies from compliance- to deterrence driven in 1994 and the INAIL's acquired responsibility to improve OHS standards with compliance-driven policies and grants since the early 2000s. This has happened due to the separation of powers doctrine enforced by the Constitution and the Italian criminal justice system model, policies and traditions. In other words, with a less regulated political system, Italian OHS enforcement policies might be much more similar to the British one. Hence, these findings highlight how discretionary practices in concert with political and criminal justice system policies, practices and traditions are responsible for decriminalisation of OHS crimes and contribute to the erosion of values of equality and justice.

Penalties and charges deter businesses from breaching the regulation and help to increased compliance culture before the regulator's visit. In other words, on-the-spot penalties or charges, improve the deterrence power of regulatory enforcement institutions and thus the efficiency of the enforcement activities. The FFI system introduced in Britain in 2012 has increased the deterrence-driven enforcement policies of the HSE, which has contributed to increasing the criminalisation of OHS crimes and the redistributive power of the regulation.

Enforcement officers in both jurisdictions complained that the reason why they can only impose fees in Britain and why only penalties in Italy is due to the separation of powers doctrine. The reason for the different responses but similar motivation is because the British and Italian political systems principles are different. The former system confers much more power to the government (the Executive), while the latter to the Parliament (the Legislative). Hence, in Britain the government is allowed to generate revenue from services or through, as in this case, independent regulators, but it is not allowed to penalise and criminalise citizens, which is the Judiciary's role. The FFI policy, however, has the same effect of penalties, which is probably the reason why HMRC has imposed a £11 million cap on the amount of money collected through fees by the HSE annually. In Italy, any charge imposed by the regulator must be codified through acts of Parliament, but the intimate relationship between UPGs and the Judiciary Authority also allows officers to issue on-the-spot penalties. The money collected through charges and penalties, however, must go straight to the Treasury. There is an explicit recognition from both jurisdictions enforcement officers that the level of fees or penalties firms are subjected to pay must be scrutinised and controlled. Hence, penalties and charges have the capacity to change significantly the level of criminalisation of OHS crimes, are effective means to achieve compliance effectively and efficiently and can help to achieve the redistributive goals of the regulation.

6   Final Remarks

From this critical comparative analysis, it can be concluded that the OHS regulation is important to achieve modern liberal values of equality and justice. The enforcement policies adopted to achieve the regulatory goals are key because law enforcement institutions and officers' legal powers are essential for achieving these. The traditional deterrence-driven law enforcement strategies have achieved results in many different criminal justice system contexts in the last two centuries, and the formulation and implementation of innovative regulatory enforcement policies since the 1970s can definitely be considered valuable in literature and policy terms. One of the greatest issues with these innovative compliance-driven enforcement policies, however, is that they tend to reduce the importance of the primary role that economic regulations have, which is to pragmatically repair the social problems caused by the adverse effects of the modern capitalist industrialisation process. In other words, economic regulations, and particularly the OHS one in this context, are social redistributive tools. They redistribute wealth and justice, and thus promote equality, and preserve and promote citizens' civil rights. The inadequate redistributive power of the OHS regulation and the under-criminalisation of these crimes is proof of an ineffective regulation and of an economic regulatory system. The literature suggesting alternative compliance-driven enforcement policies without considering and attempting to repair for the under-criminalisation issue, thus, are not helping to achieve justice. Indeed, the literature does not decide whether the state implements these policies, and this critical comparative analysis has attempted to analyse and explain why and how OHS crimes are under-criminalised in practice. The reasons, indeed, are intricate, and often the results of complex mediated political and social policies and traditions. These, however, seem to be affected significantly by global economic interests.

In simple terms, economic regulations represent an obstacle to the accumulation of capital and economic success of the national economy. The pragmatic response of regulations to the problem caused by the capitalist mode of production, indeed, cost money and harms workers. This issue is exacerbated by the increasing growth of private capital and the consequent power it has acquired globally together with the shrinking role of the state characteristic of a neoliberal economic model. While the growth and increasing movement of private capital represents a flourishing global phenomenon, nation-states' social redistributive policies and regulations are still largely constrained by national borders. This is problematic because while private capital can compete almost unconstrained across the globe, the spread of regulations and people's civil rights are limited by national borders. Political rhetoric and TINA mantras suggest that this global economic phenomenon forces national economies to reform in order to create the most attractive environment for investments, but as a consequence citizens' civil rights and social redistributive policies are eroded. The bourgeoisie and capital, indeed, also benefit from the regulation and other redistributive social policies, but the importance of achieving economic surpluses is becoming ever more important than preserving basic principles of justice and equality. Rather than using the economy as a means to promote social welfare, capital's surplus is ever more becoming an end in itself. Hence, the under-criminalisation of OHS crimes represents a failure to achieve social justice and equality. This is a fundamental political and social problem because the drive to economic surplus and nation-state's economic success depends on its capabilities to maintain economic growth while preserving citizens' civil rights through social redistributive policies.

The state's responsibility to ensure basic civil rights, however, is only guaranteed through people's votes and social struggles, at least in Britain where there is no written constitution that regulates in detail the state–citizen relationship. It seems that the economic success of nation-states and the executive power required to remain economically and politically competitive globally is more important than ensuring civil rights, democratic participation, and social equality and justice. The Human Rights Act 1998, which is the first comprehensive statute making the British state accountable for citizens' human rights, has partially repaired this issue (Sanders and Young 2007), but this is not a constitutional law and, thus, can be reformed more easily than constitutional statutes. In other words, without codified constitutional principles aiming at ensuring citizens' civil rights, the state does not have to be consistently accountable to its citizens. Without a constitution, political rhetoric supporting capitalist interests and TINA mantras, which are eroding citizens' civil rights, can become more important and urgent than ensuring equality and justice. The under-criminalisation of OHS crimes and the enforcement policy reforms introduced in Britain since the early 2000s is proof of this. The HSE's unwillingness to be scrutinised by independent academics, which was experienced during the course of the fieldwork informing these findings, and other previous research studies, is also proof of the regulator and state's lack of transparency and accountability.

The adverse effects of an expanding global capital and economic efficiency has partially been prevented in Italy by the Constitution (Caracciolo 2015), which encourages, promotes and guarantees basic modern principles of equality and justice and thus safeguards citizens' civil rights. The Constitution guarantees all citizens' civil rights, including their right to political representation, justice and equality. This is way OHS crimes in Italy do not seem to be under-criminalised at regulator level as much as in Britain. In addition, the regulatory redistributive role of INAIL is in part established by Constitutional principles, which means that this institution cannot be abolished easily. The fiscal weight, or burden, imposed on businesses as a consequence of this compulsory insurance is becoming ever more pressing for an economy that struggles to compete with emerging unregulated and, hence, competitive economic markets. The implementation of Italian prepositive constitutional principles is not completed, and it is a constant work in progress because continually interacting with emerging ideologies and trends (Calamandrei 1955). Italian constitutional values have increasingly been threatened by global capitalist economic interests (Malleson 2016). The Italian Constitution promotes collectivist Keynesian principles and hence significantly limits the capacity to implement neoliberal policies (Zagrebelsky 2013). The conflictual relationship between global capital and workers' rights is increasingly imbalanced by the state's growing bias towards capitalist interests, which is also transpiring through an endemic crisis of political transparency and accountability. The strict separation of powers the Italian Constitution envisages between political institutions and citizens is forbidding, or at least preventing, the Italian state to support these interests fully. These are probably the reasons why the Italian Constitution is claimed to be in crisis. Since the 1980s, there have been increasing attempts to reform the Constitution, at times successfully, by attempting to relax the state's institutional separation of powers and to improve governability.[1] The improvement of governability, however, seems to be intimately related to the capability of creating economic surplus. In can be argued, therefore, that the Italian Constitution is not in crisis, but in an

---

[1]One of the latest constitutional reforms proposed by Prime Minister Matteo Renzi (Democratic Party), for example, was to abolish citizens' direct election of upper house representatives (Senators) and instead allow regional governments to nominate them. This proposal, which was voted against by the Italian citizens in the 2016 constitutional referendum, would have reduced citizens' democratic decisions.

ideological conflict with political and economic ideologies that are constantly and persistently eroding workers and citizens' rights. In fact, the problem is not the governability deficit caused by the Italian Constitution, but constitutional principles granting citizens' rights, ensuring institutional separation of powers and preventing conflicts of interest, which run counter to global economic trends and interests.

References

Bardach, E., & Kagan, R. A. (1982). *Going by the book: The problem of regulatory unreasonableness*. Philadelphia: Temple University Press.

Bartrip, P. W. J., & Fenn, P. (1980a). The conventionalization of factory crime a re-assessment. *International Journal of the Sociology of Law, 8*, 175–186.

Bartrip, P. W. J., & Fenn, P. (1980b) The administration of safety: The enforcement policy of the early factory inspectorate 1844–1864, *Public Administration, 58*(Spring), 87–102.

Bartrip, P. W. J., & Fenn, P. (1983). The evolution of regulatory style in the nineteenth century British factory inspectorate. *Journal of Law and Society, 10*(1983), 201–222.

Becker, G. S. (1968). Crime and punishment: An economic approach. *Journal of Political Economy, 76*(2), 169–182.

Black, J. (2001). *Managing discretion.* Paper presented at the ARLC conference [Online]. Available from: http://www.lse.ac.uk/collections/law/staff%20 publications%20full%20text/black/alrc%20managing%20discretion.pdf. Accessed 25 Aug 2014.

Braithwaite, J. (1987). Negotiation versus litigation: Industry regulation in Great Britain and the United States. *American Bar Foundation Research Journal, 2*(1987), 559–574.

Braithwaite, J. (1997). On speaking softly and carrying big sticks: Neglected dimensions of a republican separation of powers. *University of Toronto Law Journal, 47*(3), 305–361.

Calamandrei, P. (1955, January 26). *Discorso sulla Costitutzione.* Speech delivered to the Milan Societa Umanitaria. Available from: http://www.napoliassise.it/costituzione/discorsosullacostituzione.pdf. Accessed 7 Mar 2018.

Caracciolo, L. (2015). *La Costituzione nella palude: Indagine su trattati al di sotto di ogni sospetto.* Imprimatur: Reggio Emilia.

Carson, W. G. (1970a, October). White collar crime and the enforcement of factory legislation, *British Journal of Criminology, 10*(4), 383–398.

Carson, W. G. (1970b). Some sociological aspects of strict liability and the enforcement of factory legislation. *Modern Law Review, 33,* 396.

Carson, W. G. (1979). The conventionalisation of early factory crime. *International Journal of the Sociology of Law, 7,* 37–60.

Centre for Corporate Accountability (CCA). (2008). *Fines against most companies convicted following work-related deaths less than 1/700th of their turnover, new research shows* [Online]. Available from: http://www.corporateaccountability.org.uk/press_releases/2008/mar16sent.htm. Accessed 25 Aug 2014.

Council of Europe. (2010, December 17–18). *European standards as regards the independence of the judicial system: Part II—The prosecution service.* Study N° 494/2008, CDL-AD(2010)040, January 3. Adopted by the European Commission for Democracy Through Law at its 85th Plenary Session (Venice) [Online]. Available from: http://www.venice.coe.int/webforms/documents/?pdf=CDL-AD(2010)040-e. Accessed 20 Mar 2018.

Damaška, M. R. (1973). Evidentiary barriers to conviction and two models of criminal procedure: A comparative study. *University of Pennsylvania Law Review, 506,* 1972–1973.

Damaška, M. R. (1986). *The faces of justice and state authority: A comparative approach to the legal process.* New Haven and London: Yale University Press.

Delle Fave, C. (2013). *Manuale di polizia giudiziaria.* Santarcangelo di Romagna: Maggiole Editore.

de Secondat, C.-L., & Baron de La Brède et de Montesquieu. (2001). *The spirit of the laws* (Translated from the French, by D. W. Carrithers & T. Nugent). Kitchener, ON: Batoche Books (Originally printed in 1748).

Dubini, R. (2001). *Articolo 2087 del codice civile. L'obbligo del datore di lavoro di attenersi al principio della massima sicurezza tecnologicamente fattibile.* Sicurezza tecnica, organizzativa e procedural [Online]. Available from: http://www.dbworld.it/file/studi/2087_1329822805.pdf. Accessed 25 Aug 2014.

Dworkin, R. (1977). *Taking rights seriously.* Cambridge: Duckworth Press.

Edwards v. National Coal Board. (1949). 1 All ER 743.

English Oxford Dictionary. (2017). Discretion. In *The English Oxford dictionary* [Online]. Available from: https://en.oxforddictionaries.com/definition/discretion. Accessed 17 Aug 2017.

European Commission. (2007). Improving quality at work: Community strategy 2007–2012 on health and safety at work. Communication from the Commission to the European Parliament, the Council, the European

Economic and Social Committee and the Committee of the Regions. COM(2007) 62 [Online]. Available from: http://eur-lex.europa.eu/legal-content/EN/TXT/?uri=celex%3A52007DC0062. Accessed 30 Jan 2018.

European Commission. (2012). Enforcement of fundamental workers' rights. Directorate-General for Internal Policies of the Union. ISBN: 978-92-823-3831-5. https://doi.org/10.2861/1781 [Online]. Available from: https://publications.europa.eu/en/publication-detail/-/publication/2b47fb86-73eb-4354-b5ef-996c3d461e25. Accessed 30 Jan 2018.

Fioravanti, M. (2011). Le dottrine dello stato e della costituzione. In R. Romanelli (Ed.), *Storia dello Stato Italiano dall'unità ad Oggi*. Donzelli: Italy.

Fooks, G. (2008). *The relationship between the levels of fines imposed upon companies convicted of health and safety offences resulting from deaths, and the turnover and gross profits of these companies*. Centre for Corporate Accountability [Online]. Available from: http://www.corporateaccountability.org.uk/dl/manslaughter/reform/ccasentresearchmar08.doc. Accessed 25 Aug 2014.

Galligan, D. J. (1986). *Discretionary powers*. Oxford: OUP.

Gustapane, A. (2012). *Il ruolo del pubblico ministero nella Costituzione italiana*. Bologna: Bononia University Press.

Hawkins, K. (1984). *Environment and enforcement: Regulation and the social definition of pollution*. Oxford: Clarendon Press.

Hawkins, K. (2002). *Law as last resort: Prosecution decision-making in a regulatory agency*. Oxford: Oxford University Press.

HSE. (2015). HSE *enforcement policy statement* [Online]. Available from: http://www.hse.gov.uk/pubns/hse41.pdf. Accessed 31 Jan 2018.

Hutter, B. (1997). *Compliance: Regulation and environment*. Oxford: Clarendon Press.

Langbein, J. H., Lerner, R. L., & Smith, B. P. (2009). *History of the common law: The development of Anglo-American legal institutions*. New York: Aspen Publishers. ISBN 978-0-7355-6290-5.

Malleson, T. (2016). *Fired up about capitalism, between the lines*. Toronto: Between the Lines.

Mousourakis, G. (2015). *Roman law and the origins of the civil law tradition*. Basel: Springer International Publishing. ISBN 978-3-319-12267-0; e-ISBN 978-3-319-12268-7; https://doi.org/10.1007/978-3-319-12268-7.

Nelken, D. (2010). *Comparative criminal justice: Making sense of difference*. London (UK), Thousand Oaks (CA), New Delhi (IND) and Singapore: SAGE Publications Ltd.

Ogus, A. (1994). *Regulation: Legal form and economic theory*. Oxford: Clarendon Press.

Packer, H. L. (1968). *The limits of criminal sanction*. Stanford, CA: Stanford University Press.

Pais, P. R. (2008). *Nuova normativa di tutela e salute sui luoghi di lavoro*. Roma: Epc.

Porreca, G. (2008). *Istituita con Il D. Lgs. n. 124/2004 La Diffida E La Prescrizione Obbligatoria Per Gli Ispettori Del Lavoro. Ma E' Stata Fatta Chiarezza Sull'attivita' Di P. G. In Materia Di Sicurezza Sul Lavoro?* [Online]. Available from: http://www.porreca.it/Presentazione%20decreto%20funzioni%20ispettive.htm. Accessed 2 Oct 2008.

Rausei, P (2006). *Codice delle ispezioni Volume 1 e Volume 2*. Italy: Kluwer Italia.

Rausei, P. (2008). *Vigilanza, ispezioni e sanzioni. La nuova disciplina*. Italy: Wolters Kluwer.

Repubblica Italiana. (1994). *Decreto Legislativo n. 626, 19 settembre 1994*. Attuazione delle direttive 89/391/CEE, 89/654/CEE, 89/655/CEE, 89/656/CEE, 90/269/CEE, 90/270/CEE, 90/394/CEE, 90/679/CEE, 93/88/CEE, 95/63/CE, 97/42/CE, 98/24/CE, 99/38/CE, 99/92/CE, 2001/45/CE, 2003/10/CE, 2003/18/CE e 2004/40/CE riguardanti il miglioramento della sicurezza e della salute dei lavoratori durante il lavoro. *Gazzetta Ufficiale* n.265 del 12 novembre 1994. Supplemento Ordinario n. 141.

Rinaldi, M. (2012). *Il procedimento ispettivo*. Italia: Giuffrè Ediotore.

Sanders, A., & Young, R. (2007). *Criminal justice* (3rd ed.). Oxford: Oxford University Press.

Slapper, G. (2000). *Blood in the bank: Social and legal aspects of death at work*. Aldershot: Ashgate.

Snider, L. (1991). The regulatory dance: Understanding reform processes in corporate crime. *International Journal of Sociology of Law, 19*(2), 209–237.

Taylor, A. J. (1972). *Laissez-faire and state intervention in nineteenth-century Britain*. London: Palgrave Macmillan.

Tombs, S., & Whyte, D. (2008). *A crisis of enforcement: The decriminalisation of death and injury at work* (pp. 1746–6938). London: Centre for Crime and Justice Studies [Online]. Available from: http://www.crimeandjustice.org.uk/sites/crimeandjustice.org.uk/files/crisisenforcementweb.pdf. Accessed 25 Aug 2014.

Tombs, S., & Whyte, D. (2010). *Regulatory surrender: Death, injury and the non-enforcement of law*. Liverpool: Institute of Employment Rights.

Vile, M. J. C. (1963). *Constitutionalism and Separation of Powers*. Oxford: Clarendon Press.

Warwick, P. (2006). *Policy Horizons and Parliamentary Government*. Basingstoke, Hampshire: Palgrave Macmillan.

Zagrebelsky, G. (2013). *Fondata sul lavoro. La solitudine dell'articolo 1*. Bologna: Einaudi.

Bibliography

Almond, P. (2007, July). Regulation crisis: Evaluating the potential legitimizing effects of corporate manslaughter cases. *Law & Policy, 29*(3), 285–310.

Almond, P. (2008, June). Investigating health and safety regulation: Finding room for small-scale projects [Special Issue: Law's reality: Case studies in empirical research on law]. *Journal of Law and Society, 35*(Suppl. 1), 108–125.

Almond, P. (2009). The dangers of hanging baskets: Regulatory myths and media representations of health and safety regulation. *Journal of Law and Society, 34*(3), 352–375. https://www.jstor.org/stable/25621978?seq=1#page_scan_tab_contents.

Almond, P. (2013). *Corporate manslaughter and regulatory reform.* Basingstoke: Palgrave Macmillan.

Andrews, P. (2007). Are market failure analysis and impact assessment useful? In S. Weatherill (Ed.), *Better regulation* (pp. 49–82). Portland: Hart.

Ayres, I., & Braithwaite, J. (1992). *Responsive regulation: Transcending the deregulation debate.* New York: Oxford University Press.

Baldwin, R. (1990). Why rules don't work. *The Modern Law Review, 53*(May), 321–332.

Baldwin, R. (1995). *Rules and government.* Oxford: Oxford University Press.

Baldwin, R., & Black, J. (2008). Really responsive regulation. *Modern Law Review, 71*(1), 59–94.

© The Editor(s) (if applicable) and The Author(s) 2019
D. Canciani, *The Politics and Practice of Occupational Health and Safety Law Enforcement,* Critical Criminological Perspectives, https://doi.org/10.1007/978-3-319-98509-1

Baldwin, R., & Hawkins, K. (1984). Discretionary justice: Davis reconsidered. *Public Law, 580*(Winter), 570–599.

Baldwin, R., & Veljanovski, C. G. (1984). Regulation by cost-benefit analysis. *Public Administration, 62*(Spring), 51–69.

Ball, S. J. (1994). Political interviews and the politics of interviewing. In G. Walford (Ed.), *Researching the powerful in education.* London: UCL Press.

Bardach, E., & Kagan, R. A. (1982). *Going by the book: The problem of regulatory unreasonableness.* Philadelphia: Temple University Press.

Bartrip, P. W. J., & Fenn, P. (1980a). The conventionalization of factory crime a re-assessment. *International Journal of the Sociology of Law, 8,* 175–186.

Bartrip, P. W. J, & Fenn, P. (1980b). The administration of safety: The enforcement policy of the early factory inspectorate 1844–1864. *Public Administration, 58*(Spring), 87–102.

Bartrip, P. W. J., & Fenn, P. (1983). The evolution of regulatory style in the nineteenth century British factory inspectorate. *Journal of Law and Society, 10*(1983), 201–222.

Becker, G. S. (1968). Crime and punishment: An economic approach. *Journal of Political Economy, 76*(2), 169–182.

Becker, H. S. (1963). *Outsiders: Studies in the sociology of deviance.* New York: Free Press.

Becker, H. S. (Ed.). (1964). *The other side. Perspectives on deviance.* New York: Free Press.

Bernard, R. H. (2013). *Social research method: Qualitative and quantitative approaches* (2nd ed.). Thousand Oaks, CA: Sage.

Bernstein, M. (1955). *Regulating business by independent commission.* Princeton: Princeton University Press.

Bibbings, R. (2014). *Twenty four arguments for increased action by the Health and Safety Commission and Executive.* Royal Society for the Prevention of Accidents [Online]. Available from: http://www.rospa.com/drivertraining/morr/info/24arguments.pdf. Accessed 25 Aug 2014.

Black, J. (1997). *Rules and regulators.* Oxford: Oxford University Press.

Black, J. (2001). *Managing discretion.* Paper presented at the ARLC conference [Online]. Available from: http://www.lse.ac.uk/collections/law/staff%20publications%20full%20text/black/alrc%20managing%20discretion.pdf. Accessed 25 Aug 2014.

Black, J., & Baldwin, R. (2010). Really responsive risk-based regulation. *Law & Policy, 32*(2), 181–213.

Bonomi, B., & Marinaro, M. (2009). *Dossier sicurezza. Salute e sicurezza sui luoghi di lavoro: Le strategie di prevenzione degli infortuni sul lavoro e di promozione dei livelli di salute e sicurezza sul lavoro.* Carsoli: Ministro del Lavoro, della Salute e delle Politiche Sociali and il Sole 24 S.p.A [Online]. Available from: http://www.lavoro.gov.it/SicurezzaLavoro/Documents/Dossier_sicurezza_web_EXE.pdf. Accessed 25 Aug 2014.

Bourdieu, P. (1998). *Practical reason: On the theory of action.* Stratford: Stratford University Press.

Braithwaite, J. (1985). *To punish or persuade: Enforcement of coal mine safety.* Albany: State University of New York Press.

Braithwaite, J. (1987). Negotiation versus litigation: Industry regulation in Great Britain and the United States. *American Bar Foundation Research Journal, 2*(1987), 559–574.

Braithwaite, J. (1993). Responsive business regulatory institutions. In C. Cody & C. Sampford (Eds.), *Business, ethics and law.* Sydney: Federation Press.

Braithwaite, J. (1997). On speaking softly and carrying big sticks: Neglected dimensions of a republican separation of powers. *University of Toronto Law Journal, 47*(3), 305–361.

Braithwaite, J. (2002). Rules and principles: A theory of legal certainty. *Australian Journal of Legal Philosophy, 27,* 47–82.

Braithwaite, J. (2003). What's wrong with the sociology of punishment? *Theoretical Criminology, 7*(1), 5–28.

Braithwaite, J., & Drahos, P. (2002). Zero tolerance, naming and shaming: Is there a case for it with crimes of the powerful? *Australian and New Zealand Journal of Criminology, 35*(3), 269–288.

Broadfoot, E. M., Osborn, M. J., Gilly, M., & Blucher, A. (1993). *Perception's of teaching. Teachers' lives in England and France.* London: Cassells.

Broadfoot, P. (2000). Interviewing in a crosscultural context: Some issues for comparative research. In S. Hillyard (Series ed.). *Studies in qualitative methodology.* Volume 6, pp. 53–65. London: Emerald Group Publishing Ltd.

Bronitt, S., & Stenning, P. (2011). Understanding discretion in modern policing. *Criminal Law Journal, 35*(6), 319–330.

Bryman, A. (1988). *Quantity and quality in social research.* London: Unwin Hyman.

Bryman, A. (2008). *Social research method* (3rd ed.). Oxford: Oxford University Press.

Cabinet Office News Release. (2006, January 11). *New bill to enable delivery of swift and efficient regulatory reform to cut red tape: Jim Murphy.* CAB/001 [Online]. Available from: http://webarchive.nationalarchives.gov. uk/20071001194811/bre.berr.gov.uk/regulation/news/2006/060111.asp. Accessed 25 Aug 2014.

Calamandrei, P. (1955, January 26). *Discorso sulla Costitutzione.* Speech delivered to the Milan Societa Umanitaria. Available from: http://www.napoliassise.it/costituzione/discorsosullacostituzione.pdf. Accessed 7 Mar 2018.

Caracciolo, L. (2015). *La Costituzione nella palude: Indagine su trattati al di sotto di ogni sospetto.* Imprimatur: Reggio Emilia.

Carson, W. G. (1970a, October). White collar crime and the enforcement of factory legislation. *British Journal of Criminology, 10*(4), 383–398.

Carson, W. G. (1970b). Some sociological aspects of strict liability and the enforcement of factory legislation. *Modern Law Review, 33,* 396.

Carson, W. G. (1979). The conventionalisation of early factory crime. *International Journal of the Sociology of Law, 7,* 37–60.

Carson, W. G. (1982). *The other price of Britain's oil.* Oxford: Martin Robertson.

Cataldi, E. (1983). *L'Istituto Nazionale per l'Assicurazione contro gli Infortuni sul Lavoro: testimonianza di un secolo,* INAIL 1983 [Online]. Available from: http://www.inail.it/internet/default/INAILcomunica/ListaPubblicazioni/p/ DettaglioPubblicazioni/index.html?wlpnewPage_contentDataFile=UCM_ PORTSTG_104030&wlpnewPage__dettaglioDaArchivio=true&_windowLabel=newPage. Accessed 25 Aug 2014.

Centre for Corporate Accountability (CCA). (2008). *Fines against most companies convicted following work-related deaths less than 1/700th of their turnover, new research shows* [Online]. Available from: http://www.corporateaccountability.org.uk/press_releases/2008/mar16sent.htm. Accessed 25 Aug 2014.

Centre for Occupational and Environmental Health. (2013). *The health and occupation research network* [Online]. Available from: http://www.population-health.manchester.ac.uk/epidemiology/COEH/research/thor/. Accessed 25 Aug 2014.

Cherry, N., Meyer, J. D., Adisesh, A., Brooke, R., Owen-Smith, V., Swales, C., et al. (2000). Surveillance of occupational skin disease: EPIDERM and OPRA. *The British Journal of Dermatology, 142*(6), 1128–1134.

Cicourel, A. V. (1968). *The social organization of juvenile justice.* New York: Wiley.

Coffee, J. C. (1981, January). No soul to damn: No body to kick. An unscandalized inquiry into the problem of corporate punishment. *Michigan Law Review, 79*(3), 386–459.

Cole, G. F., Frankowski, S., & Gertz, M. G. (1987). *Major criminal justice systems: A comparative survey*. Beverly Hills: Sage.

Coordinamento Tecnico Interregionale. (2011). *Attività delle regioni e delle province autonome per la prevenzione nei luoghi di lavoro, elaborazione PREO as reported by Conferenza delle regioni e delle provincie autonome* [Online]. Available from: http://www.regione.emilia-romagna.it/sicurezza-nei-luoghi-di-lavoro/coordinamento/altre-strutture-e-documenti-di-riferimento/piani-nazionali-e-regionali. Accessed 25 Aug 2014.

Copi, I. M., & Cohen, C. (1994). *Introduction to logic*. New York: Macmillan.

Council of Europe. (2010, December 17–18). *European standards as regards the independence of the judicial system: Part II—The prosecution service*. Study N° 494/2008, CDL-AD(2010)040, January 3. Adopted by the European Commission For Democracy Through Law at its 85th plenary session (Venice) [Online]. Available from: http://www.venice.coe.int/webforms/documents/?pdf=CDL-AD(2010)040-e. Accessed 20 Mar 2018.

Council of the European Union. (1989, July 29). *EEC council directive 89/391/EEC*. Introducing measures to encourage improvements in the health and safety of workers at work. *Official Journal, L 183*, 1–8.

Cowling, M. (2011). Can Marxism make sense of crime? *Global Discourse, 2*(2), 59–74.

Cranston, R. (1979). *Regulating business*. London: Macmillan.

Damaška, M. R. (1973). Evidentiary barriers to conviction and two models of criminal procedure: A comparative study. *University of Pennsylvania Law Review, 506*, 1972–1973.

Damaška, M. R. (1986). *The faces of justice and state authority: A comparative approach to the legal process*. New Haven and London: Yale University Press.

Dammer, H. R., & Albanese, J. S. (2011). *Comparative criminal justice*. Belmont: Thomson.

Dantzker, M. L., & Hunter, R. (2012). *Research methods for criminology and criminal justice* (3rd ed.). Sudbury: Jones & Bartlett Learning.

Davis, C. (2004). *Making companies safe: What works?* Centre for Corporate Accountability [Online]. Available from: http://www.unitetheunion.org/uploaded/documents/Making%20Companies%20Safe%20-%20what%20works%20(CCA-Unite%20paper)11-4856.pdf. Accessed 25 Aug 2014.

Davis, M., Croall, H., & Tyrer, J. (2010). *Criminal justice* (4th ed.). Harlow: Pearson Education Limited.

Dawson, S., Willman, P., Clinton, A., & Bamford, M. (1988). *Safety at work: The limits of self-regulation*. Cambridge: Cambridge University Press.

Delle Fave, C. (2013). *Manuale di polizia giudiziaria*. Santarcangelo di Romagna: Maggiole Editore.

de Secondat, C.-L., & Baron de La Brède et de Montesquieu. (2001). *The spirit of the laws*. (Translated from the French, by D. W. Carrithers & T. Nugent). Kitchener, ON: Batoche Books (Originally printed in 1748).

Department of the Environment, Transport and the Regions (DETR). (2000). *Revitalising health and safety*. Wetherby: Department of the Environment, Transport and the Regions [Online]. http://www.hse.gov.uk/statistics/pdf/prog2009.pdf. Accessed 25 Aug 2014.

Department of Work and Pensions. (2011a). *Good health and safety, good for everyone* [Online]. Available from: https://www.gov.uk/government/publications/good-health-and-safety-good-for-everyone. Accessed 25 Aug 2014.

Department for Work and Pensions. (2011b, November 28). *Reclaiming health and safety for all: Professor Löfstedt's independent review of health and safety legislation and the government response* [Online]. Available from: https://www.gov.uk/government/uploads/system/uploads/attachment_data/file/66790/lofstedt-report.pdf. Accessed 25 Aug 2014.

Deutscher, I. (1973). Asking questions cross-culturally: Some problems of linguistic comparability. In D. Warwick & S. Osherson (Eds.), *Comparative research methods*. Englewood Cliffs, NJ: Prentice Hall.

Diver, C. S. (1980). A theory of regulatory enforcement. *Public Policy, 28*(3), 257–299.

Dubini, R. (2001). *Articolo 2087 del codice civile. L'obbligo del datore di lavoro di attenersi al principio della massima sicurezza tecnologicamente fattibile*. Sicurezza tecnica, organizzativa e procedural [Online]. Available from: http://www.dbworld.it/file/studi/2087_1329822805.pdf. Accessed 25 Aug 2014.

Dugan, E. (2013, July 30). Compensation culture is a myth: Claims for work-related injuries and diseases fall 60 per cent in a decade. *The Independent* [Online]. Available from: http://www.independent.co.uk/news/uk/home-news/compensation-culture-is-a-myth-claims-for-workrelated-injuries-and-diseases-fall-60-per-cent-in-a-decade-8738679.html. Accessed 25 Aug 2014.

Dworkin, R. (1977). *Taking rights seriously*. Cambridge: Duckworth Press.

DWP. (2016). *Framework document between the Health and Safety Executive and the Department of Work and Pension* [Online]. Available from: http://data.parliament.uk/DepositedPapers/Files/DEP2016-0071/DWP_HSE_FRAMEWORK_DOCUMENT_2016.pdf. Accessed 11 Apr 2018.

Edwards v. National Coal Board. (1949). 1 All ER 743.

Emerson, R. M. (1969). *Judging delinquents: Context and process in juvenile court*. Chicago: Aldine.

English Oxford Dictionary. (2017). Discretion. In *The English Oxford Dictionary* [Online]. Available from: https://en.oxforddictionaries.com/definition/discretion. Accessed 17 Aug 2017.

European Commission. (2005). *Labour inspection (health and safety) in the EU (25 member states)—A short guide*. DG Employment, Social Affair and Equal Opportunities [Online]. Available from: http://ec.europa.eu/employment_social/health_safety/slic_en.htm. Accessed 18 Nov 2008.

European Commission. (2007). *Improving quality at work: Community strategy 2007–2012 on health and safety at work*. Communication from the Commission to the European Parliament, the Council, the European Economic and Social Committee and the Committee of the Regions. COM(2007) 62 [Online]. Available from: http://eur-lex.europa.eu/legal-content/EN/TXT/?uri=celex%3A52007DC0062. Accessed 30 Jan 2018.

European Commission. (2012). *Tax burden on labour* [Online]. Available from: http://ec.europa.eu/europe2020/pdf/themes/20_tax_burden_on_labour.pdf. Accessed 25 Aug 2014.

European Commission. (2012b). *Enforcement of fundamental workers' rights*. Directorate-General for Internal Policies of the Union. ISBN: 978-92-823-3831-5. https://doi.org/10.2861/1781 [Online]. Available from: https://publications.europa.eu/en/publication-detail/-/publication/2b47fb86-73eb-4354-b5ef-996c3d461e25. Accessed 30 Jan 2018.

Fields, C. B., & Moore, R. H. (Eds.). (2005). *Comparative criminal justice*. Prospect Heights, IL: Waveland Press.

Fioravanti, M. (2011). Le dottrine dello stato e della costituzione. In R. Romanelli (Ed.), *Storia dello Stato Italiano dall'unità ad Oggi*. Donzelli: Italy.

Fisse, B. (1983, December). Reconstructing corporate criminal law: Deterrence, retribution, fault, and sanctions. *Southern California Law Review, 56*(6), 1141–1246.

Fitz, J., & Halpin, D. (1994). Ministers and mandarins: Educational research in elite settings. In G. Walford (Ed.), *Researching the powerful in education*. London: UCL Press.

Flick, U. (2009). *An introduction to qualitative research* (4th ed.). London: Sage.

Fooks, G. (2008). *The relationship between the levels of fines imposed upon companies convicted of health and safety offences resulting from deaths, and the turnover and gross profits of these companies.* Centre for Corporate Accountability [Online]. Available from: http://www.corporateaccountability.org.uk/dl/manslaughter/reform/ccasentresearchmar08.doc. Accessed 25 Aug 2014.

Fooks, G., Bergman D., & Rigby, B. (2007). *International comparison of (a) techniques used by state bodies to obtain compliance with health and safety law and accountability for administrative and criminal offences and (b) sentences for criminal offences.* Health and Safety Executive [Online]. Available from: http://www.hse.gov.uk/research/rrhtm/rr607.htm. Accessed 25 Aug 2014.

Frey, L., Botan, C., & Kreps, G. (1999). *Investigating communication: An introduction to research methods* (2nd ed.). Boston: Allyn & Bacon.

Galligan, D. J. (1986). *Discretionary powers.* Oxford: Oxford University Press.

Gert, B. (1998). *Morality: Its nature and justification.* Oxford: Oxford University Press.

Gewirtz, S., & Ozga, J. (1994). Interviewing the education policy elite. In G. Walford (Ed.), *Researching the powerful in education.* London: UCL Press.

Glaser, B., & Strauss, A. (1967). *The discovery of grounded theory: Strategies for qualitative research.* New York: Aldine.

Gobert, J., & Punch, M. (2003). *Rethinking corporate crime.* London: Butterworths.

Goffman, E. (1959). *The presentation of self in everyday life.* Garden City: Doubleday Anchor.

Goffman, E. (1961). *Encounters: Two studies in the sociology of interaction.* Indianapolis: Bobbs-Merrill.

Goffman, E. (1963). *Behavior in public places: Notes on the social organization of gatherings.* New York: Free Press.

Goffman, E. (1967). *Interaction ritual: Essays on face-to-face behaviour.* New York: Anchor Books.

Goffman, E. (1970). *Strategic interaction.* Oxford: Blackwell.

Goffman, E. (1971). *Relations in public: Microstudies of the public order.* Harmondsworth: Penguin.

Grabosky, P., & Braithwaite, J. (1986). *Of manners gentle: Enforcement strategies of Australian business regulatory agencies.* Melbourne: Oxford University Press.

Gray, D. (2013). *Doing research in the real world* (3rd ed.). London: Sage.

Gray, W., & Scholz, T. (1991). Analysing the equity and efficiency of OSHA enforcement. *Law & Policy, 13*(3), 185–214.

Gray, W. B., & Scholz, J. T. (1993). Does regulatory enforcement work? A panel analysis of OSHA enforcement. *Law & Society Review, 27*(1), 177–185.

Great Britain. (1974). *Health and Safety at Work etc. Act 1974 (c. 37).* London: HMSO.

Great Britain. (1995). *The Report of Injuries, Diseases and Dangerous Occurrences Regulation 1995.* London: HMSO.

Great Britain. (1998). *Health and Safety (Enforcing Authority) Regulations 1998 (c. 494).* London: HMSO.

Great Britain. (2007). *Corporate Manslaughter and Corporate Homicide Act 2007 (c. 19).* London: HMSO.

Great Britain. (2008). *Health and Safety (Offence) Act 2008 (c. 20).* London: HMSO.

Great Britain. (2012). *Health and Safety (Fees) Regulation 2012 (c. 255).* London: HMSO.

Great Britain. (2013). *Reporting of Injuries, Diseases and Dangerous Occurrences Regulation 2013* (No. 1471). Norwich: The Stationary Office [Online]. Available from: http://www.legislation.gov.uk/uksi/2013/1471/contents/made. Accessed 23 Mar 2017.

Great Britain, House of Lord Select Committee on Constitution Eleventh Report. (2006). *Legislative and Regulatory Reform Bill* [Online]. Available from: http://www.publications.parliament.uk/pa/ld200506/ldselect/ldconst/194/19403.htm#n3. Accessed 25 Aug 2014.

Guba, E. G., & Lincoln, Y. S. (1994). Competing paradigms in qualitative research. In N. K. Denzin & Y. S. Lincoln (Eds.), *Handbook of qualitative research.* London: Sage.

Gunningham, N. (1987). Negotiated non-compliance: A case study of regulatory failure. *Law and Policy, 9*(1), 69–82.

Gunningham, N. (2007). Corporate environmental responsibility: Law and the limits of voluntarism. In D. McBarnet, A. Voiculescu, & T. Campbell (Eds.), *The new corporate accountability: Corporate social responsibility and the law.* Cambridge: University Press.

Gunnigham, N., & Gabrowsky, P. (1998). *Smart regulation: Designing environmental policy.* Oxford: Clarendon Press.

Gunningham, N., & Johnstone, R. (1999). *Regulating workplace safety: System and sanctions.* Oxford: Oxford University Press.

Gustapane, A. (2012). *Il ruolo del pubblico ministero nella Costituzione italiana.* Bologna: Bononia University Press.

Guthrie, C., Rachlinski, J. J., & Wistrich, A. J. (2001). Inside the judicial mind. *Cornell Law Review, 86,* 777–790.

Gye, H. (2013, December 4). No more elf and safety: Government launches crackdown on 'bonkers' bans on traditional Christmas fun. *Daily Mail* [Online]. Available from: http://www.dailymail.co.uk/news/article-2517997/Government-crackdown-bonkers-Christmas-bans.html. Accessed 25 Aug 2014.

Hampton, P. (2005, March). *Reducing administrative burden: Effective inspections and enforcement.* HM Treasury [Online]. Available from: http://www.fera.defra.gov.uk/aboutUs/betterRegulation/documents/hamptonPrinciples.pdf. Accessed 25 Aug 2014.

Hantrais, L., & Mangan, S. (1996). *Cross-national research methods in the social sciences.* London: Pinter.

Harvey, L. (1990). *Critical social research.* London: Unwin Hyman.

Hawkins, K. (1991). Enforcing regulation: More of the same from pearce and tombs. *British Journal of Criminology, 31*(4), 427–430.

Hawkins, K. (1983). Bargain and bluff: Compliance strategy and deterrence in the enforcement of regulation. *Law & Policy Quarterly, 5*(1), 35–73.

Hawkins, K. (1984). *Environment and enforcement: Regulation and the social definition of pollution.* Oxford: Clarendon Press.

Hawkins, K. (1989, July). 'FATCATS' and prosecution in a regulatory agency: A footnote on the social construction of risk. *Law & Policy, 11*(3), 370–391.

Hawkins, K. (1992a). *The uses of discretion.* Oxford: Oxford University Press.

Hawkins, K. (1992b). The use of legal discretion: Perspectives from law and social science. In K. Hawkins (Ed.), *The uses of discretion.* Oxford: Oxford University Press.

Hawkins, K. (1992c). *The regulation if occupational health and safety: A socio-legal perspective.* London: Health and Safety Executive.

Hawkins, K. (2002). *Law as last resort: Prosecution decision-making in a regulatory agency.* Oxford: Oxford University Press.

Hawkins, K., & Thomas, J. (Eds.). (1984a). *Enforcing regulation.* Boston: Kluwer-Nijhoff.

Hawkins, K., & Thomas, J. (1984b). The enforcement process in regulatory bureaucracies. In K. Hawkins & J. Thomas (Eds.), *Enforcing regulation.* Boston: Kluwer-Nijhoff.

Hazards Magazine. (2005, August). Protection racket. *Hazards Magazine,* Issue 91 [Online]. Available from: http://www.hazards.org/commissionimpossible/protectionracket.htm. Accessed 25 Aug 2014.

Hazards Magazine. (2007, January–March). Safety repressed. *Hazards Magazine,* Issue 97 [Online]. Available from: http://www.hazards.org/safetyreps/safetyrepressed.htm. Accessed 25 Aug 2014.

Hazards Magazine. (2010, October–December). Get shirty. *Hazards Magazine,* Issue 112 [Online]. Available from: http://www.hazards.org/votetodie/getshirty.htm. Accessed 25 Aug 2014.

Hazards Magazine. (2011, October–December). Overkill. *Hazards Magazine,* Issue 116 [Online]. Available from: http://www.hazards.org/votetodie/overkill.htm. Accessed 25 Aug 2014.

Hazards Magazine. (2013, April–June). Robbed! Bloody bandages but no bloody compensation. *Hazards Magazine,* Issue 122 [Online]. Available from: http://www.hazards.org/votetodie/robbed.htm. Accessed 25 Aug 2014.

Hazards Magazine. (2015). *Give up. What can you do when a watchdog just sucks?* [Online]. Available from: http://www.hazards.org/votetodie/giveup.htm. Accessed 15 June 2015.

Health and Safety Commission. (1999). *Highlights from the HSC annual report and the HSC/E accounts 1998/1999* [Online]. Available from: http://www.hse.gov.uk/aboutus/reports/index.htm. Accessed 25 Aug 2014.

Hill, A. (2006, November 22). Memorial service for Aberfan disaster. *The Observer* [Online]. Available at: http://observer.guardian.co.uk/uk_news/story/0,,1928511,00.html. Accessed 25 Aug 2014.

Hopkins, A. (1995). *Making safety work: Getting management commitment to occupational health and safety.* Sydney: Allen & Unwin.

House of Commons Work and Pensions Select Committee. (2004). *The work of the Health and Safety Commission and Executive.* Fourth Report of Session 2003–04, Vol. I, HC 456–1. London: The Stationery Office [Online]. Available from: http://www.publications.parliament.uk/pa/cm200304/cmselect/cmworpen/456/45602.htm. Accessed 25 Aug 2014.

HSC. (2000a). *Highlights from the HSC annual report and the HSC/E accounts 1999/2000* [Online]. Available from: http://www.hse.gov.uk/aboutus/reports/index.htm. Accessed 25 Aug 2014.

HSC. (2000b). *Health and safety statistics 1999/2000* [Online]. Available from: http://www.hse.gov.uk/statistics/pdf/hss9900.pdf?pdf=hss9900. Accessed 10 Apr 2017.

HSC. (2001). *Highlights from the HSC annual report and the HSC/E accounts 2000/2001* [Online]. Available from: http://www.hse.gov.uk/aboutus/ reports/index.htm. Accessed 25 Aug 2014.

HSC. (2002). *Highlights from the HSC annual report and the HSC/E accounts 2001/2002* [Online]. Available from: http://www.hse.gov.uk/aboutus/ reports/index.htm. Accessed 25 Aug 2014.

HSC. (2003). *Highlights from the HSC annual report and the HSC/E accounts 2002/2003* [Online]. Available from: http://www.hse.gov.uk/aboutus/ reports/index.htm. Accessed 25 Aug 2014.

HSC. (2004a, April 6). *Becoming a modern regulator.* Health and Safety Commission Paper HSC/04/53 [Online]. Available from: http://www.hse. gov.uk/aboutus/meetings/hscarchive/2004/060404/c53.pdf. Accessed 25 Aug 2014.

HSC. (2004b). *Highlights from the HSC annual report and the HSC/E accounts 2003/2004* [Online]. Available from: http://www.hse.gov.uk/aboutus/ reports/index.htm. Accessed 25 Aug 2014.

HSC. (2005a). *Sensible health and safety at work: The regulatory methods used in Great Britain.* An account of the approach of the Health and Safety Commission [Online]. Available from: http://www.hse.gov.uk/aboutus/ strategiesandplans/sensiblehealthandsafety.pdf. Accessed 25 Aug 2014.

HSC. (2005b). *Highlights from the HSC annual report and the HSC/E accounts 2004/2005* [Online]. Available from: http://www.hse.gov.uk/aboutus/ reports/index.htm. Accessed 25 Aug 2014.

HSC. (2006). *Highlights from the HSC annual report and the HSC/E accounts 2005/2006* [Online]. Available from: http://www.hse.gov.uk/aboutus/ reports/index.htm. Accessed 25 Aug 2014.

HSC. (2007). *Highlights from the HSC annual report and the HSC/E accounts 2006/2007.* Available from: http://www.hse.gov.uk/aboutus/reports/index. htm. Accessed 25 Aug 2014.

HSC. (2008). *Highlights from the HSC annual report and the HSC/E accounts 2007/2008* [Online]. Available from: http://www.hse.gov.uk/aboutus/ reports/index.htm. Accessed 25 Aug 2014.

HSE. (1998). *Health and Safety (Enforcing Authority) Regulations 1998: A–Z guide to allocation* [Online]. Available from: http://www.hse.gov.uk/foi/ internalops/og/og-00073-appendix1.htm. Accessed 18 June 2013.

HSE. (2003a). *Driving at work, managing work-related road safety* [Online]. Available from: http://www.hse.gov.uk/pubns/indg382.pdf. Accessed 25 Aug 2014.

HSE. (2003b). *Regulation, enforcement, inspection and what we will do.* Paper presented to the HSE board, October [Online]. Available from: http://www.hse.gov.uk/aboutus/meetings/hsearchive/2003/030903/item7.pdf. Accessed 25 Aug 2014.

HSE. (2004). *A strategy for workplace health and safety in Britain to 2010 and beyond* [Online]. Available from: http://www.hse.gov.uk/aboutus/strategie-sandplans/strategy2010.pdf. Accessed 25 Aug 2014.

HSE. (2009a). *HSE annual report and accounts 2008/2009* [Online]. Available from: http://www.hse.gov.uk/aboutus/reports/index.htm. Accessed 25 Aug 2014.

HSE. (2009b). *Enforcement policy statement* [Online]. Available from: http://www.hse.gov.uk/pubns/hse41.pdf. Accessed 25 Aug 2014.

HSE. (2010). *HSE annual report and accounts 2009/2010* [Online]. Available from: http://www.hse.gov.uk/aboutus/reports/index.htm. Accessed 25 Aug 2014.

HSE. (2011). *HSE annual report and accounts 2010/2011* [Online]. Available from: http://www.hse.gov.uk/aboutus/reports/index.htm. Accessed 25 Aug 2014.

HSE. (2012a). *HSE annual report and accounts 2011/2012* [Online]. Available from: http://www.hse.gov.uk/aboutus/reports/index.htm. Accessed 25 Aug 2014.

HSE. (2012b). *Historical picture. Trends in work-related injuries and ill health since the introduction of the Health and Safety at Work Act (HSWA) 1974* [Online]. Available from: http://www.hse.gov.uk/statistics/history/histori-cal-picture.pdf. Accessed 25 Aug 2014.

HSE. (2012c). *Fee for intervention: What you need to know* [Online]. Available from: http://www.hse.gov.uk/pubns/hse48.pdf. Accessed 10 Dec 2012.

HSE. (2013a). *HSC/HSE merger enforcement statement* [Online]. Available from: http://www.hse.gov.uk/aboutus/furtherinfo/merger.htm. Accessed 25 Aug 2014.

HSE. (2013b). *Local authority enforcement* [Online]. Available from: http://www.hse.gov.uk/lau/enforcement.htm. Accessed 25 Aug 2014.

HSE. (2013c). *How HSE enforces health and safety* [Online]. Available from: http://www.hse.gov.uk/enforce/enforce.htm#enfpen. Accessed 25 Aug 2014.

HSE. (2013d). *Data source* [Online]. Available from: http://www.hse.gov.uk/statistics/sources.htm#employment. Accessed 25 Aug 2014.

HSE. (2013e). *HSE annual report and accounts 2012/2013* [Online]. Available from: http://www.hse.gov.uk/aboutus/reports/index.htm. Accessed 25 Aug 2014.

HSE. (2013f). *Fees for intervention* [Online]. Available from: http://www.hse.gov.uk/fee-for-intervention/. Accessed 25 Aug 2014.

HSE. (2013g). *Health and safety statistics* [Online]. Available from: http://www.hse.gov.uk/statistics/index.htm. Accessed 25 Aug 2014.

HSE. (2013h). *Enforcement management model* [Online]. Available from: http://www.hse.gov.uk/enforce/emm.pdf. Accessed 25 Aug 2014.

HSE. (2013i). *National local authority enforcement code health and safety at work England, Scotland & Wales* [Online]. Available from: http://www.hse.gov.uk/lau/national-la-code.pdf. Accessed 23 Feb 2018.

HSE. (2013j). *Reporting accidents and incidents at work. A brief guide to the Reporting of Injuries, Diseases and Dangerous Occurrences Regulations 2013 (RIDDOR)* [Online]. Available from: http://www.hse.gov.uk/pubns/indg453.pdf. Accessed 23 Mar 2017.

HSE. (2013k). *Health and safety executive statistics 2009/10* [Online]. Available from: http://www.hse.gov.uk/statistics/overall/hssh0910.pdf?pdf=hssh0910. Accessed 10 Apr 2010.

HSE. (2014a). *Penalties* [Online]. Available from: http://www.hse.gov.uk/enforce/enforcementguide/court/sentencing-penalties.htm. Accessed 25 Aug 2014.

HSE. (2014b). *How HSE meets the obligations in the statutory regulators' compliance code* [Online]. Available from: http://www.hse.gov.uk/regulation/compliancecode/. Accessed 25 Aug 2014.

HSE. (2014c). *HSE and Local Authority (LA) regulators working together* [Online]. Available from: http://www.hse.gov.uk/laU/index.htm. Accessed 25 Aug 2014.

HSE. (2014d). *ALARP "at a glance"* [Online]. Available from: http://www.hse.gov.uk/Risk/theory/alarpglance.htm. Accessed 25 Aug 2014.

HSE. (2014e). *Media centre* [Online]. Available from: http://press.hse.gov.uk/. Accessed 25 Aug 2014.

HSE. (2014f). *HSE annual report 2013/14* [Online]. Available from: http://www.hse.gov.uk/aboutus/reports/1314/ar1314.pdf. Accessed 3 Mar 2017.

HSE. (2015a). *Busting the health and safety myths* [Online]. Available from: http://www.hse.gov.uk/myth/. Accessed 9 June 2015.

HSE. (2015b). *HSE principles for Cost Benefit Analysis (CBA) in support of ALARP decisions* [Online]. Available from: http://www.hse.gov.uk/risk/theory/alarpcba.htm. Accessed 15 June 2015.

HSE. (2015c). *Cost Benefit Analysis (CBA) checklist* [Online]. Available from: http://www.hse.gov.uk/risk/theory/alarpcheck.htm. Accessed 15 June 2015.

HSE. (2015d). HSE *enforcement policy statement* [Online]. Available from: http://www.hse.gov.uk/pubns/hse41.pdf. Accessed 31 Jan 2018.

HSE. (2015e). *HSE annual report 2014/15* [Online]. Available from: http://www.hse.gov.uk/aboutus/reports/ara-2014-15.pdf. Accessed 3 Mar 2017.

HSE. (2016). *HSE annual report 2015/16* [Online]. Available from: https://www.gov.uk/government/uploads/system/uploads/attachment_data/file/534093/hse-annual-report-and-accounts-2015-2016.pdf. Accessed 3 Mar 2017.

HSE. (2017a). *HSE annual report and accounts 2016/17* [Online]. Available from: http://www.hse.gov.uk/aboutus/reports/ara-2016-17.pdf. Accessed 3 Mar 2017.

HSE. (2017b). *Health and safety at work summary statistics for Great Britain 2017* [Online]. Available from: http://www.hse.gov.uk/statistics/overall/hssh1617.pdf. Accessed 10 Apr 2017.

HSE. (2018a). *Local authority enforcement* [Online]. Available from: http://www.hse.gov.uk/lau/enforcement.htm. Accessed 23 Feb 2018.

HSE. (2018b). *How HSE regulates* [Online]. Available from: http://www.hse.gov.uk/enforce/. Accessed 23 Feb 2018.

HSE. (2018c). *Advice and guidance* [Online]. Available from: http://www.hse.gov.uk/enforce/advice-information-guidance.htm. Accessed 23 Feb 2018.

HSE. (2018d). *Is health surveillance required in my workplace?* [Online]. Available from: http://www.hse.gov.uk/health-surveillance/requirement/index.htm. Accessed 23 Mar 2018.

Huntington, S. (1952). The Marasmus of the ICC: The commission, the railroads, and the public interest. *Yale Law Journal, 614,* 467–509.

Hutter, B. (1988). *The reasonable arm of the law? The law enforcement procedures of environmental health officers.* Oxford: Clarendon Press.

Hutter, B. (1997). *Compliance: Regulation and environment.* Oxford: Clarendon Press.

Hutter, B., & Manning, P. (1990). The contexts of regulation: The impact upon health and safety inspectorates in Britain. *Law and Policy, 12*(2), 103–136.

Impresanews. (2013, May 6). INAIL, esauriti in pochi secondi 155 milioni di contributi. Oltre 13 mila domande arrivate dall imprese di tutta Italia. Impresanews.it [Online]. Available from: http://www.impresanews.it/News/Agevolazioni-finanziarie/363/inail--esauriti-in-pochi-secondi-155-milioni-di-contributi. Accessed 25 Aug 2014.

INAIL. (2003). *Bilancio Consuntivo 2002* [Online]. Available from: https://www.inail.it/cs/internet/docs/1_bil-cons-2002-pdf.pdf. Accessed 6 Apr 2018.

INAIL. (2007). *Bilancio Consuntivo 2006* [Online]. Available from: https://www.inail.it/cs/internet/docs/bilancio-cons-2006-pdf.pdf. Accessed 6 Apr 2018.

INAIL. (2014a). *La storia* [Online]. Available from: http://www.inail.it/internet/default/Chisiamo/Lastoria/index.html. Accessed 25 Aug 2014.

INAIL. (2014b). *Banca Dati* [Online]. Available from: http://bancadaticsa.inail.it/bancadaticsa/login.asp. Accessed 25 Aug 2014.

INAIL. (2014c). *Bilancio Consuntivo 2013* [Online]. Available from: https://www.inail.it/cs/internet/docs/all-bilancio-consuntivo-2013.pdf. Accessed 6 Apr 2018.

INAIL. (2016). *Bilancio Consuntivo 2015* [Online]. Available from: https://www.inail.it/cs/internet/docs/ammt-bilancio-consuntivo-2015.pdf. Accessed 6 Apr 2018.

INAIL. (2017a). *Relazione annuale Inail: nel 2016 flessione degli infortuni mortali sul lavoro* [Online]. Available from: https://www.inail.it/cs/internet/comunicazione/sala-stampa/comunicati-stampa/com-stampa-relazione-annuale-inail-2016.html. Accessed 29 Jan 2018.

INAIL. (2017b). *Bilancio Previsione 2017* [Online]. Available from: https://www.inail.it/cs/internet/docs/ammt-bilancio-previsione-2017.pdf. Accessed 6 Apr 2018.

International Businesses Organisations USA. (2009). *Italy: Justice system and national police handbook. Volume 1 strategic and criminal justice system.* Washington, DC: International Business Publications.

Johnson, J. (2010). Beating fatigue. *RoSPA Occupational Safety & Health Journal, 40*(11), 27–30.

Jones, S. (2013). *Criminology.* Oxford: Oxford University Press.

Jost, P. J. (1997). Regulatory enforcement in the presence of a court system. *International Review of Law and Economics, 17,* 491–508.

Justice Committee. (2009). *The crown prosecution service: Gatekeeper of the criminal justice system.* Session 2008–09, 9th Report. London: Justice Committee publications [Online]. Available from: http://www.publications.parliament.uk/pa/cm200809/cmselect/cmjust/186/18602.htm. Accessed 25 Aug 2014.

Kagan, R. (1978). *Regulatory justice: Implementing a wage price freeze.* New York: Russell Sage Foundation.

Kagan, R. A., & Scholz, J. T. (1984). The criminology of the corporation and regulatory enforcement strategies. In K. Hawkins & J. Thomas (Eds.), *Enforcing regulation* (pp. 67–96). Hingham, MA: Kluwer-Nijhoff.

Katzen, S. (2006, May). Cost-benefit analysis: Where should we go from here? *Fordham Urban Law Journal, 33*(4), 1313–1317.

Kirk, J., & Miller, M. (1986). *Reliability and validity in qualitative research* (Qualitative Research Method Series, Vol. 1). London: Sage.

Kogan, M. (1994). Researching the powerful in education and elsewhere. In G. Walford (Ed.), *Researching the powerful in education*. London: UCL Press.

Laffont, J. J., & Tirole, J. (1991). The politics of government decision making: A theory of regulatory capture. *Quarterly Journal of Economics, 106*(4), 1089–1127.

Langbein, J. H., Lerner, R. L., & Smith, B. P. (2009). *History of the common law: The development of Anglo-American legal institutions*. New York: Aspen Publishers. ISBN 978-0-7355-6290-5.

Levine, M. E., & Forrence, J. L. (1990). Regulatory capture, public interest, and the public agenda: Toward a synthesis. *Journal of Law Economics & Organization, 6*, 167–198.

Lincoln, Y. S., & Guba, E. G. (1985). *Naturalistic inquiry*. Beverley Hills, CA: Sage.

Lord Robens. (1972). *Report of the committee on safety and health at work 1970–1972*. London: HMSO.

Lord Young of Graffham. (2010, October 15). *Common sense, common safety: A report by Lord Young of Graffham: A report by Lord Young of Graffham following a Whitehall-wide review of the operation of health and safety laws*. The Cabinet Office Policy Paper [Online]. Available from: https://www.gov.uk/government/uploads/system/uploads/attachment_data/file/60905/402906_CommonSense_acc.pdf. Accessed 25 Aug 2014.

Ma, Y. (2002). Prosecutorial discretion and plea bargaining in the United States, France, Germany, and Italy: A comparative perspective. *International Criminal Justice Review, 12*, 22–52.

Makkai, T., & Braithwaite, J. (1994). Reintegrative shaming and regulatory compliance. *Criminology, 32*, 361–385.

Makkai, T., & Braithwaite, J. (1991). Criminological theories and regulatory compliance. *Criminology, 29*, 191–220.

Malleson, T. (2016). *Fired up about capitalism, between the lines*. Toronto: Between the Lines.

Manning, P. K. (1977). *Police work: The social organization of policing*. Cambridge: MIT Press.

McBarnet, D., Voiculescu, A., & Campbell, T. (Eds.). (2007). *The new corporate accountability: Corporate social responsibility and the law.* Cambridge: University Press.

Meindinger, E. (1986). Regulatory culture: A theoretical outline. *Law and Policy, 9,* 355–386.

Messina, D. (2008). *Salviamo la costituzione italiana: Il tema che dominerà la nuova stagione politica.* Milano: RCS Libry S.p.A.

Mickelson, R. A. (1994). A feminist approach to researching the powerful in education. In G. Walford (Ed.), *Researching the powerful in education.* London: UCL Press.

Ministero della Salute: Direzione generale della programmazione sanitaria, dei livelli di assistenza e dei principi etici di sistema. (2004). *Rapporto nazionale di monitoraggio dei livelli essenziali di assistenza anno 2001* [Online]. Available from: http://www.salute.gov.it/imgs/C_17_pubblicazioni_1175_allegato.pdf. Accessed 25 Aug 2014.

Ministero della Salute: Direzione generale della programmazione sanitaria, dei livelli di assistenza e dei principi etici di sistema. (2006). *Rapporto nazionale di monitoraggio dei livelli essenziali di assistenza anni 2002–2003* [Online]. Available from: http://www.salute.gov.it/imgs/C_17_pubblicazioni_1173_allegato.pdf. Accessed 25 Aug 2014.

Ministero della Salute: Direzione generale della programmazione sanitaria, dei livelli di assistenza e dei principi etici di sistema. (2007). *Rapporto nazionale di monitoraggio dei livelli essenziali di assistenza anno 2004* [Online]. Available from: http://www.salute.gov.it/imgs/C_17_pubblicazioni_1174_allegato.pdf. Accessed 25 Aug 2014.

Ministero della Salute: Direzione generale della programmazione sanitaria, dei livelli di assistenza e dei principi etici di sistema. (2009). *Rapporto nazionale di monitoraggio dei livelli essenziali di assistenza anni 2005–2006* [Online]. Available from: http://www.salute.gov.it/imgs/C_17_pubblicazioni_1072_allegato.pdf. Accessed 25 Aug 2014.

Ministero della Salute: Direzione generale della programmazione sanitaria, dei livelli di assistenza e dei principi etici di sistema. (2010). *Rapporto nazionale di monitoraggio dei livelli essenziali di assistenza anni 2007–2009* [Online]. Available from: http://www.salute.gov.it/imgs/C_17_pubblicazioni_1674_allegato.pdf. Accessed 25 Aug 2014.

Ministero del Lavoro e delle Politiche Sociali. (2004). *Circolare Ministro del Lavoro e delle Politiche Sociali n. 24 del 24 giugno 2004* [Online]. http://www.inps.it/circolariZip/Circolare%20numero%20132%20del%2020-9-2004_Allegato%20n%201.pdf. Accessed 25 Aug 2014.

Ministro Del Lavoro, Della Salute Delle Politiche Sociali. (2008). Decreto 18 settembre 2008 [Online]. Available from: http://www.gazzettaufficiale.it/atto/serie_generale/caricaDettaglioAtto/originario?atto.dataPubblicazioneGazzetta=2008-10-18&atto.codiceRedazionale=08A07492&elenco-30giorni=false. Accessed 30 Jan 2018.

Ministry of Housing, Communities & Local Government. (2011). *Annex A8: Revenue outturn cultural, environmental and planning services (RO5) 2009–10* (revised) [Online]. Available from: https://www.gov.uk/government/uploads/system/uploads/attachment_data/file/560117/Annex_A8.xls. Accessed 3 Apr 2018.

Ministry of Housing, Communities & Local Government. (2012). *Annex A8: Revenue outturn cultural, environmental and planning services (RO5) 2010–11* [Online]. Available from: https://www.gov.uk/government/uploads/system/uploads/attachment_data/file/15252/2123447.xls. Accessed 3 Apr 2018.

Ministry of Housing, Communities & Local Government. (2013). *Annex A8: Revenue outturn cultural, environmental, regulatory and planning services (RO5) 2011 to 2012* [Online]. Available from: https://www.gov.uk/government/uploads/system/uploads/attachment_data/file/15356/revenue_outturn_2011-12_final_annex_a8.xls. Accessed 3 Apr 2018.

Ministry of Housing, Communities & Local Government. (2014). *Annex A8: Revenue outturn cultural, environmental, regulatory and planning services (RO5) 2012 to 2013* [Online]. Available from: https://www.gov.uk/government/uploads/system/uploads/attachment_data/file/282502/Annex_A8_Revenue_Outturn_Cultural__Environmental__Regulatory_and_Planning_Services__RO5__2012-13__revised_.xls. Accessed 3 Apr 2018.

Ministry of Housing, Communities & Local Government. (2015). *Annex A8: Revenue outturn cultural, environmental, regulatory and planning services (RO5) 2013 to 2014* [Online]. Available from: https://www.gov.uk/government/uploads/system/uploads/attachment_data/file/379849/Annex_A8_-_Revenue_Outturn_Cultural__Environmental__Regulatory_and_Planning_Services__RO5__2013-14.xlsx. Accessed 3 Apr 2018.

Ministry of Housing, Communities & Local Government. (2016). *Revenue outturn (RO5) data 2014 to 2015 by local authority* (revised) [Online]. Available from: https://www.gov.uk/government/uploads/system/uploads/attachment_data/file/497101/Revenue_Outturn__RO5__data_2014-15_by_LA_-_02-Feb-2016.xls. Accessed 3 Apr 2018.

Ministry of Housing, Communities & Local Government. (2017a). *Revenue outturn cultural, environmental, regulatory and planning services (RO5) 2015 to 2016* [Online]. Available from: https://www.gov.uk/government/uploads/system/uploads/attachment_data/file/659793/RO5_2015-16_data_by_LA_-_Revision.xlsx. Accessed 3 Apr 2018.

Ministry of Housing, Communities & Local Government. (2017b). *Revenue outturn cultural, environmental, regulatory and planning services (RO5) 2016 to 2017* [Online]. Available from: https://www.gov.uk/government/uploads/system/uploads/attachment_data/file/659778/RO5_2016-17_data_by_LA.xlsx. Accessed 3 Apr 2018.

Morse, J. M. (1998). Designing funded qualitative research. In N. Denzin & Y. S. Lincoln (Eds.), *Strategy of qualitative research*. London: Sage.

Mousourakis, G. (2015). *Roman law and the origins of the civil law tradition*. Basel: Springer International Publishing. ISBN 978-3-319-12267-0; e-ISBN 978-3-319-12268-7; https://doi.org/10.1007/978-3-319-12268-7.

Mumola, C. (199, January). Substance abuse and treatment, state and federal prisoners, 1997. *Bureau of Justice Statistics Special Report*. U.S. Department of Justice. Office of Justice Programs [Online]. Available from: http://www.iapsonline.com/sites/default/files/Substance%20Abuse%20and%20Treatment%20of%20State%20and%20Federal%20Prisoners,%201997.pdf. Accessed 25 Aug 2014.

Nelken, D. (2010). *Comparative criminal justice: Making sense of difference*. London (UK), Thousand Oaks (CA), New Delhi (IND) and Singapore: SAGE Publications Ltd.

Neuman, W. L. (2014). *Basics of social research: Qualitative and quantitative approaches*. 3rd Edition. Harlow (UK): Paerson Education Ltd.

Newburn, T., & Reiner, R. (2007). Policing and the police. In M. Maguire, R. Morgan, & R. Reiner (Eds.), *The Oxford handbook of criminology* (4th ed.). Oxford: Oxford University Press.

Nichols, T., & Armstrong, P. (1973). *Safety or profit: Industrial accidents and conventional wisdom*. Bristol: Falling Wall Press.

Ogus, A. (1994). *Regulation: Legal form and economic theory*. Oxford: Clarendon Press.

Olsen, P. (1992). *Six cultures of regulation*. Copenhagen: Handelshojskolen.

Packer, H. L. (1968). *The limits of criminal sanction*. Stanford, CA: Stanford University Press.

Pais, P. R. (2008). *Nuova normativa di tutela e salute sui luoghi di lavoro*. Roma: Epc.

Pearce, F. (1976). *Crimes of the powerful. Marxism, crimes and deviance*. London: Pluto Press.

Pearce, D. W. (1983). *Cost-benefit analysis*. London: Macmillan.

Pearce, F., & Tombs, S. (1990). Ideology, hegemony and empiricism: Compliance theories of regulation. *British Journal of Criminology, 30*(4), 423–443.

Pearce, F., & Tombs, S. (1991). Policing corporate 'skid rows': A reply to Keith Hawkins. *British Journal of Criminology, 31*(4), 415–426.

Pearce, F., & Tombs, S. (1998). Foucault, governmentality, Marxism. *Social & Legal Studies, 7*(4), 567–575.

Pelliccia, L. (2008). *Il nuovo testo unico di sicurezza sul lavoro*. Rimini: Maggioli Editore.

Petroni, A. M. (2004, July 7). *L'analisi costi/benefici ed i suoi riflessi sul sistema politico ed amministrativo*. Paper presented at the Conferenza annuale della Ragioneria Generale dello Stato in Rome [Online]. Available from: http://www.astrid-online.it/Economia-e/Studi--ric/Archivio-2/Petroni_Rag_Stato_29_07_04.pdf. Accessed 25 Aug 2014.

Pollitt, C. (2004). *Unbundled government: A critical analysis of the global trend to agencies, quangos and contractualisation*. London: Routledge Studies in Public Management.

Porreca, G. (2008). *Istituita con Il D. Lgs. n. 124/2004 La Diffida E La Prescrizione Obbligatoria Per Gli Ispettori Del Lavoro. Ma E' Stata Fatta Chiarezza Sull'attivita' Di P. G. In Materia Di Sicurezza Sul Lavoro?* [Online]. Available from: http://www.porreca.it/Presentazione%20decreto%20funzioni%20ispettive.htm. Accessed 2 Oct 2008.

Posner, E. (2003, December). Transfer regulations and cost-effectiveness analysis. *Duke Law Journal, 53*(3), 1067–1079.

Posner, R. A. (1972a). *Economic analysis of law* (1st ed.). Boston: Little, Brown.

Posner, R. A. (1972b). The behaviour of administrative agencies. *The Journal of Legal Studies, 1,* 314–325.

Pruyadharshini, E. (2003). Coming unstuck: Thinking otherwise about "studying up". *Anthropology and Education Quarterly, 34*(4), 420–437.

PuntoSicuro. (2008, October 1). *Ambiente e lavoro convention: rumori e vibrazioni* [Online]. Available from: http://www.puntosicuro.it/sicurezza-sul-lavoro-C-1/tipologie-di-contenuto-C-6/informazione-formazione-addestramento-C-56/ambiente-lavoro-convention-rumori-vibrazioni-AR-8279/. Accessed 25 Aug 2014.

PuntoSicuro. (2009). Convegno in Toscana: RLS e sicurezza sul lavoro. *Puntosicuro.it,* Numero 2266, Anno 11, ottobre 23 [Online]. Available from: http://www.puntosicuro.it/sicurezza-sul-lavoro-C-1/tipologie-di-contenuto-C-6/informazione-formazione-addestramento-C-56/convegni-in-toscana-rls-sicurezza-sul-lavoro-AR-9348/. Accessed 25 Aug 2014.

Rausei, P. (2006). *Codice delle ispezioni Volume 1 e Volume 2.* Italy: Kluwer Italia.

Rausei, P., & Tiraboschi, M. (Eds.). (2014). *L'ispezione del lavoro, dieci anni dopo la riforma. Id.lgs. n. 124/2004 fra passato e future.* ADAPT Professional Series n.3. Modena: ADAPT University Press. ISBN 978-88-98652-28-0

Rausei, P. (2008). *Vigilanza, ispezioni e sanzioni. La nuova disciplina.* Italy: Wolters Kluwer.

Rawls, J. (1973). *A theory of justice.* Oxford: Oxford University Press.

Raz, J. (1972). Legal principles and the limits of law. *Yale Law Journal, 81,* 823–835.

Rees, J. (1984). Selecting strategies of control over organizational life. In K. Hawkins & J. M. Thomas (Eds.), *Enforcing regulation.* Boston: Kluwer-Nijhoff.

Rees, J. (1988). *Reforming the workplace: A study of self-regulation in occupational safety.* Philadelphia: University of Pennsylvania Press.

Regione Friuli-Venezia Giulia. (2010). *Attività UOPSAL, ASS Friuli Venezia Giulia 2000–2010* [Online]. Avialable from: http://www.regione.fvg.it/rafvg/export/sites/default/RAFVG/salute-sociale/organizzazione-salute-tutela-sociale/FOGLIA29/allegati/AttivitaSPSAL00-10xcomart27xsito.ppt. Accessed 25 Aug 2014.

Regione Friuli-Venezia Giulia, Agenzia Regionale Della Sanità. (2008). *Gli Infortuni Sul Lavoro In Friuli Venezia Giulia Atlante E Analisi Preventiva (2001–2006).* Agenzia [Online]. Available from: http://www.tdp.univ.fvg.it/sites/default/files/Studio%20%20infortuni%20sul%20lavoro%20in%20FVG%202001-2006.pdf. Accessed 25 Aug 2014.

Regione Friuli-Venezia Giulia, Azienda per i servizi Sanitari n.1 Triestina, Struttura Complessa Prevenzione E Sicurezza Negli Ambienti Di Lavoro (SCPSAL). (2013a). *Guida Utile* [Online]. Available from: http://www.ass1.sanita.fvg.it/servlet/page?_pageid=71&_dad=pass1&_schema=PASS1&act=2&id=3684. Accessed 25 Aug 2014.

Regione Friuli-Venezia Giulia, Azienda per i servizi Sanitari n.1 Triestina, Struttura Complessa Prevenzione E Sicurezza Negli Ambienti Di Lavoro (SCPSAL). (2013b). *Guida Utile: Vigilanza Negli Ambienti di Lavoro* [Online]. Available from: http://www.ass1.sanita.fvg.it/servlet/page?_pageid=71&_dad=pass1&_schema=PASS1&act=2&id=786. Accessed 25 Aug 2014.

Regione Friuli-Venezia Giulia, Azienda per i servizi Sanitari n.1 Triestina, Struttura Complessa Prevenzione E Sicurezza Negli Ambienti Di Lavoro (SCPSAL). (2013c). *Guida Utile. Direzione Amministrativa* [Online].

Available from: http://www.ass1.sanita.fvg.it/servlet/page?_pageid=71&_dad=pass1&_schema=PASS1&act=2&id=884. Accessed 25 Aug 2014.

Regione Friuli-Venezia Giulia, Azienda per i servizi Sanitari n.1 Triestina, Struttura Complessa Prevenzione E Sicurezza Negli Ambienti Di Lavoro (SCPSAL). (2013d). *Guida Utile. Direzione Strategica (Staff)* [Online]. Available from: http://www.ass1.sanita.fvg.it/servlet/page?_pageid=71&_dad=pass1&_schema=PASS1&act=2&id=2888. Accessed 25 Aug 2014.

Regione Lombardia, Dipartimento Di Prevenzione Medico, Prevenzione E Sicurezza Negli Ambienti Di Lavoro. (2013a). *Progetto Sanita' Rsa 2011/2012 – Presentazione* [Online]. Available from: http://www.asl.milano.it/ITA/Default.aspx?SEZ=2&PAG=74&NOT=5497. Accessed 25 Aug 2014.

Regione Lombardia, Dipartimento Di Prevenzione Medico, Prevenzione E Sicurezza Negli Ambienti Di Lavoro. (2013b). *Agricoltura E Manutenzione Verde* [Online]. Available from: http://www.asl.milano.it/ITA/Default.aspx?SEZ=2&PAG=74&NOT=5523. Accessed 25 Aug 2014.

Regione Lombardia, Dipartimento Di Prevenzione Medico, Prevenzione E Sicurezza Negli Ambienti Di Lavoro. (2013c). *Incontriamo i Medici Competenti* [Online]. Available from: http://www.asl.milano.it/ITA/Default. aspx?SEZ=2&PAG=74&NOT=5321. Accessed 25 Aug 2014.

Regione Marche, Giunta Regionale Servizio Sanità, Unità O. O. Sanità Pubblica e Prevenzione. (1995). *Ripercussioni Del Decreto Legislativo 758/94 Sulla Operatività Dei Servizi Di Prevenzione Nei Luoghi Di Lavoro* [Online]. Available from: cd494.mannelli.info/files/linee_guida_758.doc. Accessed 25 Aug 2014.

Regione Sicilia, Azienda Sanitaria Provinciale Trapani, Dipartimento di Prevenzione della Salute, U.O. Prevenzione igienico-sanitaria ed epidemiologia occupazionale. (2013). *Assistenza, informazione e formazione* [Online]. Available from: http://www.asptrapani.it/servizi/Menu/dinamica.aspx?idArea=18681&idCat=22333&ID=22334. Accessed 25 Aug 2014.

Regione Veneto, Azienda ULSS 13 Mirano. (2013). *Assistenza, formazione-informazione, promozione della salute* [Online]. Available from: http://www.ulss13mirano.ven.it/nqcontent.cfm?a_id=7668. Accessed 25 Aug 2014.

Reiman, J., & Leighton, P. (2010). *The rich get richer and the poor get prison: Ideology class and criminal justice.* Boston: Allyn & Bacon.

Reiss, A. J. Jr. (1971). *The police and the public.* New Haven, CT: Yale University Press.

Repubblica Italiana. (1930). *Reggio Decreto* del 19 ottobre, 1930 n.1398 (e seguenti modifiche). Codice Penale 1930. Gazzetta Ufficiale n. 251 del 26 ottobre 1930.

Repubblica Italiana. (1942). *Regio Decreto* del 16 Marzo 1942, n. 262 (e seguenti modifiche). *Codice Civile 1942.* Gazzetta Ufficiale n. 79, del 4 aprile 1942.

Repubblica Italiana. (1947). *Costituzione della Repubblica Italiana* (e seguenti modifiche). Gazzetta Ufficiale n. 298 del 27 dicembre 1947 edizione straordinaria.

Repubblica Italiana. (1955). *Decreto del Presidente della Repubblica del 27 aprile 1955, n. 520.* Riorganizzazione centrale e periferica del Ministero del lavoro e della previdenza sociale. *Gazzetta Ufficiale* n.149 del 1 luglio 1955.

Repubblica Italiana. (1965). *Decreto Del Presidente Della Repubblica n. 1124, 30 giugno 1965.* Testo unico delle disposizioni per l'assicurazione obbligatoria contro gli infortuni sul lavoro e le malattie professionali. *Gazzetta Ufficiale* n.257 del 13 ottobre 1965, Supplemento Ordinario n. 24.

Repubblica Italiana. (1978). *Legge n. 833, 23 dicembre 1978,* Istituzione del servizio sanitario nazionale. *Gazzetta Ufficiale* n. 360 del 28 dicembre 1978.

Repubblica Italiana. (1981). *Legge n. 689, 24 aprile 1981.* Legge di depenalizzazione. *Gazzetta Ufficiale* n.329 del 30 novembre 1981.

Repubblica Italiana. (1988). *Codice di Procedura Penale* (aggiornato al Decreto del Presidente della Repubblica n. 447 del 22 settembre 1988), Legge 18 giugno 1955, n. 517. *Gazzetta Ufficiale* n. 123, 15 maggio 1955.

Repubblica Italiana. (1994a). *Decreto Legislativo n. 626, 19 settembre 1994.* Attuazione delle direttive 89/391/CEE, 89/654/CEE, 89/655/CEE, 89/656/ CEE, 90/269/CEE, 90/270/CEE, 90/394/CEE, 90/679/CEE, 93/88/ CEE, 95/63/CE, 97/42/CE, 98/24/CE, 99/38/CE, 99/92/CE, 2001/45/ CE, 2003/10/CE, 2003/18/CE e 2004/40/CE riguardanti il miglioramento della sicurezza e della salute dei lavoratori durante il lavoro. *Gazzetta Ufficiale* n.265 del 12 novembre 1994. Supplemento Ordinario n. 141.

Repubblica Italiana. (1994b). *Decreto Legislativo n. 758, 19 dicembre 1994.* Modificazioni alla disciplina sanzionatoria in materia di lavoro. *Gazzetta Ufficiale* n.21 del 26 gennaio 1995, Supplemento Ordinario n. 9.

Repubblica Italiana. (1997). *Decreto Legislativo n. 281, 28 Agosto 1997.* Definizione ed ampliamento delle attribuzioni della Conferenza per i rapporti tra lo Stato, le regioni e le province autonome di Trento e Bolzano ed unificazione, per le materie ed i compiti di interesse comune delle regioni, delle province e dei comuni, con la Conferenza Stato - città ed autonomie locali. *Gazzetta Ufficiale* n. 202 del 30 agosto 1997.

Repubblica Italiana. (2000). *Decreto Legislativo n. 38, 23 febbraio 2000.* Disposizioni in materia di premi dell'Istituto Nazionale per l'Assicurazione contro gli Infortuni sul Lavoro e le malattie professionali (INAIL). *Gazzetta Ufficiale* n.50 del 1 marzo 2000.

Repubblica Italiana. (2001). *Decreto Legislativo n. 231, 8 giugno 2001.* Disciplina della responsabilità amministrativa delle persone giuridiche, delle società e delle associazioni anche prive di personalità giuridica, a norma dell'articolo 11 della legge 29 settembre 2000, n. 300. *Gazzetta Ufficiale* n.140 del 19 giugno 2001.

Repubblica Italiana. (2007). *Decreto Del Presidente Del Consiglio Dei Ministri 17 dicembre 2007.* Esecuzione dell'accordo del 1° agosto 2007, recante: "Patto per la tutela della salute e la prevenzione nei luoghi di lavoro. *Gazzetta Ufficiale* n. 3 del 4 gennaio 2007.

Repubblica Italiana. (2008). *Decreto Legislativo n. 81, 9 aprile 2008* (testo aggiornato al 15 ottobre 2010). Attuazione dell'articolo 1 della legge 3 agosto 2007, n. 123, in materia di tutela della salute e della sicurezza nei luoghi di lavoro. *Gazzetta Ufficiale* n. 101 del 30 Aprile 2008.

Repubblica Italiana. (2009). *Decreto Legislativo n. 106, 3 agosto 2009.* Disposizioni integrative e correttive del decreto legislativo 9 aprile 2008, n. 81, in materia di tutela della salute e della sicurezza nei luoghi di lavoro. *Gazzetta Ufficiale* n.180 del 5 agosto 2009, Supplemento Ordinario n. 142.

Repubblica Italiana (2010). *Legge n. 122, 30 luglio 2010.* Conversione in legge, con modificazioni, del decreto legge 31 maggio 2010, n. 78, recante misure urgenti in materia di stabilizzazione finanziaria e di competitività economica. *Gazzetta Ufficiale* n.176 del 30 luglio 2010, Supplemento Ordinario n. 174.

Repubblica Italiana. (2012). *Codice di Procedura Civile* (aggiornato al Decreto Legge 22 giugno 2012). Regio Decreto 28 ottobre 1940, n.1443. *Gazzetta Ufficiale* n. 253, 28 ottobre 1940.

Repubblica Italiana. (2015a). *Decreto Legislativo 4 marzo 2015, n. 22.* Disposizioni per il riordino della normativa in materia di ammortizzatori sociali in caso di disoccupazione involontaria e di ricollocazione dei lavoratori disoccupati, in attuazione della legge 183/2014). Pubblicato nella Gazz. Uff. 6 Marzo 2015, n. 54, S.O. [Online]. Available from: http://www.gazzettaufficiale.it/eli/id/2015/3/6/15G00036/sg. Accessed 30 Jan 2018.

Repubblica Italiana. (2015b). *Decreto Legislativo 14 settembre 2015, n. 149.* Disposizioni per la razionalizzazione e la semplificazione dell'attività ispettiva in materia di lavoro e legislazione sociale, in attuazione della legge 10 dicembre 2014, n. 183. Pubblicato nella Gazz. Uff. 23 settembre 2015, n. 221, S.O. [Online]. Available from: http://www.gazzettaufficiale.it/eli/id/2015/09/23/15G00161/sg. Accessed 30 Jan 2018.

Rhondda Cynon Taf County Borough Council. (2013). *Statement of policy reporting of injuries, diseases and dangerous occurrences regulations difference between 1985 and 1996 regulation* [Online]. Available from: http://www.rctednet.net/documents/policies/healthsafety/HS5_%20REPORTING_OF_INJURIES_DISEASES_AND_DANGEROUS_OCCURRENCES.pdf. Accessed 25 Aug 2014.

Richardson, B., Ogus, A., & Burrows, P. (1983). *Policing pollution: A study of regulation and enforcement.* Oxford: Clarendon Press.

Rideout, R. W. (1989). *Principles of labour law.* London: Sweet & Maxwell.

Rinaldi, M. (2012). *Il procedimento ispettivo.* Italia: Giuffrè Ediotore.

Ritchie, J., Spencer, L., & O'Connor, W. (2003). Carrying out qualitative analysis. In J. Ritchie & J. Lewis (Eds.), *Qualitative research practice: A guide for social science students and researchers.* London: Sage.

Ritchie, J., & Spencer, L. (1994). Qualitative data analysis for applied policy research. In A. Bryman & R. G. Burgess (Eds.), *Analyzing qualitative data.* London: Routledge.

Rogers, J. (2006). Restructuring the exercise of prosecutorial discretion in England. *Oxford Journal of Legal Studies, 26*(4), 775–803.

Ross, H. L. (1970). *Settled out of court. The social process of insurance claims adjustment.* Chicago: Aldine.

Rowan-Robinson, J., Watchman, P., & Barker, C. (1990). *Crime and regulation: A study of the enforcement of regulatory codes.* Beverley Hills, CA: T&T Clark.

Rubin, H. J., & Rubin, I. S. (2012). *Qualitative interviewing: The art of hearing data.* London: Sage.

Rubini, G. (2011, dicembre 20). Lavoro in corso. La "disposizione". *Diarioprevenzione Magazine* [Online]. Available from: http://www.diario-prevenzione.net/diarioprevenzione/html/modules.php?name=News&file=print&sid=174. Accessed 25 Aug 2014.

Ryan, G. W., & Bernard, H. R. (2003). Techniques to identify themes. *Field Methods, 15,* 85–109.

Sabine, G. H., & Thorson, T. L. (1973). *A history of political theory.* Hinsdale: Harcourt Brace.

Safety and Health Practitioner. (2013, November 4). *Businesses fined more than five million under FFI.* Safety Health Practitioner [Online]. Available from: http://www.ohs.co.uk/news/2013/11/time-to-have-an-ohs-audit-businesses-fined-more-than-five-million-under-ffi/. Accessed 25 Aug 2014.

Sanders, A., & Young, R. (2007). *Criminal justice* (3rd ed.). Oxford: Oxford University Press.

Scheele, B., & Groeben, N. (1988). *Dialog-konsenses-methoden zur rekonstruktion subjektiver theorien.* Tübingen: Francke.

Shavell, S. (1982). The social versus the private incentive to bring suit in a costly legal system. *Journal of Legal Studies, 11,* 333–339.

Shavell, S. (1999). The level of litigation: Private versus social optimality of suit and settlement. *International Review of Law & Economics, 19*, 99–115.

Shaw, I. G. R., & Gould, N. (2001). *Qualitative research in social work.* London: Sage.

Shover, N., Lynxwiler, J., Groce, S., & Clelland, D. (1984). Regional variation in regulatory law enforcement; the Surface Mining Control and Reclamation Act of 1977. In K. Hawkins & J. M. Thomas (Eds.), *Enforcing regulations.* Boston, MA: Kluwer-Nijhoff.

Silverman, D. (2011). *Interpreting qualitative data: A guide to the principles of qualitative research.* London: Sage.

Skolnick, J. H. (1966). *Justice without trial: Law enforcement in democratic society.* New York: Wiley.

Slapper, G. (2000). *Blood in the bank: Social and legal aspects of death at work.* Aldershot: Ashgate.

Slapper, G., & Kelly, D. (2013). *The English legal system: 2013–2014.* Abingdon: Routledge.

Slapper, G., & Tombs, S. (1999). *Corporate crime.* London: Longman.

Slapper, G., & Tombs, S. (2000). Corporate crime: Official statistics and the mass media. In Y. Jewkes & G. Letherby (Eds.), *Criminology: A reader.* London: Sage.

Smith, S. E. (2002, November). 'Counterblastes' to tobacco: Five decades of North American tobacco litigation. *Windsor Review of Legal and Social Issues, 14*, 1–32.

Snape, D., & Spencer, L. (2003). The foundation of qualitative research. In J. Ritchie & J. Lewis (Eds.), *Qualitative research practice: A guide for social science students and researchers.* London: Sage.

Snider, L. (1991). The regulatory dance: Understanding reform processes in corporate crime. *International Journal of Sociology of Law, 19*(2), 209–237.

Soderberg, M. (2007). Uncertainty and regulatory outcome in the Swedish electricity distribution sector European. *Journal of Law and Economics, 25*, 79–94.

Soricelli, C. (2012, dicembre 1). Morti sul lavoro. Gli ultimi dati dell'Osservatorio di Bologna. *DirittiDistorti* [Online]. Available from: http://www.dirittidistorti.it/articoli/12-lavoro/1141-morti-sul-lavoro-gli-ultimi-dati-dellosservatorio-di-bologna.html. Accessed 25 Aug 2014.

Spitzer, M. (2000). Judicial auditing. *Journal of Legal Studies, 29*, 649–683.

Steinberg, S. R., & Kincheloe, J. (2010). Power, emancipation, and complexity: Employing critical theory. *Power and Education, 2*(2), 140–151.

Stenbacka, C. (2001). Qualitative research requires quality concepts of its own. *Management Decision, 39*(7), 551–555.

Stigler, G. (1970). The optimum enforcement of law. *Journal of Political Economy, 78*(3), 526–545.

Stigler, G. (1971). The theory of economic regulation. *Bell Journal of Economics and Management Science, 2*, 3–21.

Sudnow, D. (1965). Normal crimes: Sociological features of the penal code in a public defender office. *Social Problems, 12*, 255–276.

Sung, H.-E. (2006, May) Democracy and criminal justice in cross-national perspective: From crime control to due process. *The ANNALS of the American Academy of Political and Social Science, 605*(1), pp. 311–337.

Tait, N. (2007, June 15). ECJ rejects challenge to health and safety laws. *Financial Time* [Online]. Available from: http://www.ft.com/cms/s/df854548-1add-11dc-8bf0-000b5df10621,Authorised=false.html?_i_location=http%3A%2F%2Fwww.ft.com%2Fcms%2Fs%2F0%2Fdf854548-1add-11dc-8bf0-000b5df10621.html%3Fsiteedition%3Duk&siteedition=uk&_i_referer=http%3A%2F%2Fsearch.ft.com%2Fsearch%3FqueryText%3DECJ%2Brejects%2Bchallenge%2Bto%2Bhealth%2Band%2Bsafety%2Blaws#axzz3BRR260SX. Accessed 25 Aug 2014.

Taylor, A. J. (1972). *Laissez-faire and state intervention in nineteenth-century Britain*. London: Palgrave Macmillan.

The Crown Prosecution Service. (2013). *Prosecution policy and guidance, the code for crown prosecutors* [Online]. Available from: http://www.cps.gov.uk/publications/code_for_crown_prosecutors/. Accessed 25 Aug 2014.

The Crown Prosecution Service. (2014). *Who we are* [Online]. Available from: http://www.cps.gov.uk/publications/reports/2011/who_we_are.html. Accessed 25 Aug 2014.

The Law Commission. (1998). *Consents to prosecution* [Online]. Available from: http://lawcommission.justice.gov.uk/docs/lc255_Consents_to_Prosecution.pdf. Accessed 25 Aug 2014.

The National Archives. (2013). *Enterprise and regulatory reform act 2013 explanatory notes*. London: HMSO [Online]. Available from: http://www.legislation.gov.uk/ukpga/2013/24/notes. Accessed 25 Aug 2014.

The Supreme Court. (2014). *Frequently asked questions* [Online]. Available from: http://www.supremecourt.uk/faqs.html#1a. Accessed 25 Aug 2014.

Tinti, B. (2007). *Toghe Rotte*. Roma: Chiarelettere.

Tiraboschi, M., & Fantini, L. (2009). *Il testo unico della salute e sicurezza sul lavoro dopo il correttivo (D.Lgs. n. 106/2009)*. Italia: Giuffrè Editore.

Tombs, S. (2003). Accounting for safety crimes? HSE, enforcement data and the (shifting) politics of access. *Radical Statistics, 81*(Spring), 51–61.

Tombs, S. (2016). Making better regulation, making regulation better? *Policy Studies, 37*(4), 332–349.

Tombs, S. (2017). The UK's corporate killing law: Un/fit for purpose? *Criminology and Criminal Justice* (published online on 23 August 2017). Available from: http://journals.sagepub.com/doi/full/10.1177/1748895817725559. Accessed 25 Mar 2018.

Tombs, S., & Whyte, D. (2007). *Safety crimes*. Cullompton: Willan Publishing.

Tombs, S., & Whyte, D. (2008). *A crisis of enforcement: The decriminalisation of death and injury at work* (pp. 1746–6938). London: Centre for Crime and Justice Studies [Online]. Available from: http://www.crimeandjustice.org.uk/sites/crimeandjustice.org.uk/files/crisisenforcementweb.pdf. Accessed 25 Aug 2014.

Tombs, S., & Whyte, D. (2009). A deadly consensus: Worker safety and regulatory degradation under New Labour. *British Journal of Criminology, 52*(5), 997–1016.

Tombs, S., & Whyte, D. (2010). *Regulatory surrender: Death, injury and the non-enforcement of law*. Liverpool: Institute of Employment Rights.

Tonry, M. (2008). Learning from the limitations of deterrence research. In M. Tonry (Ed.), *Crime and justice: A review of research*. Chicago: The University of Chicago Press.

Travaglio, M. (2007). Preface. In B. Tinti (Ed.), *Toghe Rotte*. Roma: Chiarelettere.

UK Statistics Authority. (2010, May). *Assessment of compliance with the code of practice for official statistics. Statistics on health and safety at work (produced by the Health and Safety Executive)*. Assessment Report 42 [Online]. Available from: http://www.hse.gov.uk/statistics/pdf/assessment-report-42.pdf. Accessed 10 Jan 2017.

UK Statistics Authority. (2013, September). *Assessment of compliance with the code of practice for official statistics. Statistics on health and safety at work (produced by the Health and Safety Executive)*. Assessment Report 261 [Online]. Available from: https://www.statisticsauthority.gov.uk/archive/assessment/assessment/assessment-reports/assessment-report-261---statistics-on-health-and-safety-at-work.pdf. Accessed 10 Jan 2017.

UNITE. (2008). *Lack of investigation 2001–2007. Incident reported to the health and safety executive* [Online]. Available from: http://www.uniteth-eunion.org/uploaded/documents/Lack%20of%20Investigation%20of%20Incidents%20reported%20to%20the%20HSE%20(Unite%20report)11-4817.pdf. Accessed 8 May 2018.

Venturi, D. (2014). Prescrizione obbligatoria—Articolo 15. In P. Rausei & M. Tiraboschi (Eds.), *L'ispezione del lavoro, dieci anni dopo la riforma. Id.lgs. n. 124/2004 fra passato e future* (ADAPT Professional Series no.3, pp. 332–343). Modena: ADAPT University Press. ISBN 978-88-98652-28-0.

Vile, M. J. C. (1963). *Constitutionalism and Separation of Powers*. Oxford: Clarendon Press.

Vogel, D. (1984). *National styles of regulation: Environmental policy in Great Britain and the United States*. Ithaca: Cornell University Press.

von Hirsch, A., Bottoms, A., Burney, E., & Wikstrom, P.-O. (1999). *Criminal deterrence and sentence severity: An analysis of recent research*. Oxford: Hart Publishing.

Walford, G. (2011). *Researching the powerful*. British Educational Research Association [Online]. Available from: http://www.bera.ac.uk/resources/resource-list?page=4&tid=All. Accessed 25 Aug 2014.

Walls, J., Pidgeon, N., Weyman, A., & Horlick-Jones, T. (2004, June). Critical trust: Understanding lay perceptions of health and safety risk regulation. *Health, Risk, and Society, 6*(2), 133–150.

Walters, D., & James, P. (1998). *Robens Revisited: The case for a review of occupational health and safety legislation*. Liverpool: Institute of Employment Rights.

Weatherill, S. (2007). *Better regulation*. Portland: Hart Publishing.

Westmarland, L. (2011). *Researching crime and justice: Tales from the field*. New York: Routledge.

Whyte, D. (2000). Researching the powerful: Towards a political economy of method. In R. King & E. Wincup (Eds.), *Doing research in crime and justice*. Oxford: Oxford University Press.

Williams, K. R., Gibbs, J. P., & Erickson, M. L. (1980). Public knowledge of statutory penalties: The extent and basis of accurate perception. *Pacific Sociological Review, 23*(1), 1980.

Woolf, A. (1973). The Robens report—The wrong approach. *Industrial Law Journal, 2*, 88–95.

Wright, B. R. E., Caspi, A., Moffitt, T. E., & Paternoster, R. (2004, May). Does the perceived risk of punishment deter criminally prone individuals?

Rational choice, self-control, and crime. *Journal of Research in Crime and Delinquency,* *41*(2), 180–213.

Yeager, P. (1991). *The limits of the law: The public regulation of private pollution.* Cambridge: Cambridge University Press.

Zagrebelsky, G. (2013). *Fondata sul lavoro. La solitudine dell'articolo 1.* Bologna: Einaudi.

Index

© The Editor(s) (if applicable) and The Author(s) 2019
D. Canciani, *The Politics and Practice of Occupational Health
and Safety Law Enforcement*, Critical Criminological Perspectives,
https://doi.org/10.1007/978-3-319-98509-1

Printed by Printforce, the Netherlands